McGraw·Hill
Learning Materials

SPECTRUM
GEOGRAPHY

COMMUNITIES

Grade 3

Authors

James F. Marran
Social Studies Chair Emeritus
New Trier Township High School
Winnetka, Illinois

Cathy L. Salter
Geography Teacher
Educational Consultant
Hartsburg, Missouri

McGraw·Hill
Learning Materials

250 Old Wilson Bridge Road
Worthington, Ohio 43085

The **McGraw·Hill** Companies

EAN

9 781577 681533

90000

Program Reviewers

Bonny Berryman
Eighth Grade Social Studies Teacher
Ramstad Middle School
Minot, North Dakota

Grace Foraker
Fourth Grade Teacher
B. B. Owen Elementary School
Lewisville Independent School District
Lewisville, Texas

Wendy M. Fries
Teacher/Visual and Performing Arts Specialist
Kings River Union School District
Tulare County, California

Maureen Maroney
Teacher
Horace Greeley I. S. 10 Queens
District 30
New York City, New York

Geraldeen Rude
Elementary Social Studies Teacher
1993 North Dakota Teacher of the Year
Minot Public Schools
Minot, North Dakota

Photo Credits

McGraw-Hill
Consumer Products

A Division of The McGraw-Hill Companies

Table of Contents

Introduction to Geography

What Is a Community?

Our Earlier Communities

Types of Communities

1

OUR PLANET EARTH

As you read about Earth, think about two of its most important features— land and water.

From space, Earth looks like a big marble. Can you guess what the blue and brown patches are? If you guessed water and land, you are right! The blue patches are water. The brown patches are land. Some of the types of land features are plains, plateaus, hills, and mountains. But that's just the beginning! There's so much more to learn about our colorful planet.

Let's start with the patches of blue—water. You can probably name many types of bodies of water. Rivers, lakes, seas, and oceans are just a few. Here's an amazing fact: nearly three fourths of Earth's surface is covered by water, but only a very small portion of that water is drinkable. Most of Earth's water is salty.

The water that surrounds the land on Earth is divided into four huge areas called oceans. They are the Pacific, Atlantic, Indian, and Arctic oceans. The largest and deepest ocean is the Pacific Ocean. **Geographers,** people who study Earth's surface, have found that the Pacific Ocean covers about one third of Earth's surface.

30% land

70% water

Let's move on to the brown patches—land. The land on Earth's surface is divided into several big pieces called **continents.** There are seven continents.

Look at the map below. See how many oceans and continents you can locate and name. Can you find all four oceans and all seven continents?

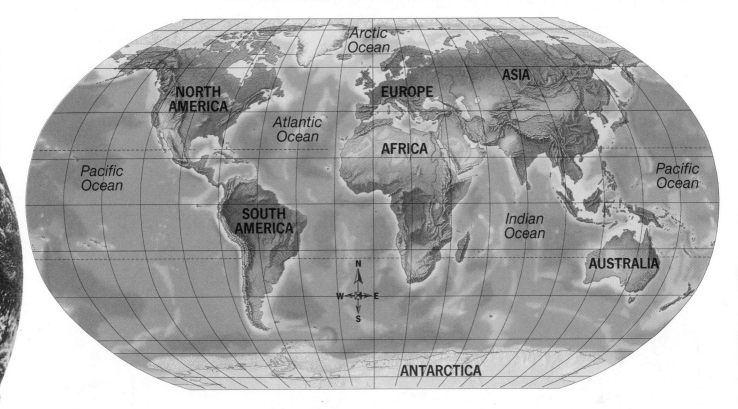

This world map of Earth's surface shows its physical features such as oceans and continents. The map is drawn on a flat surface.

A globe also shows the physical features of Earth, but it has a different shape. It is round like Earth. What continents and oceans can you see on this globe?

MAP SKILLS
Using a Physical Map to Learn about Land and Water

Two of Earth's most important physical features are land and water. This **physical map** shows some of the landforms and water forms.

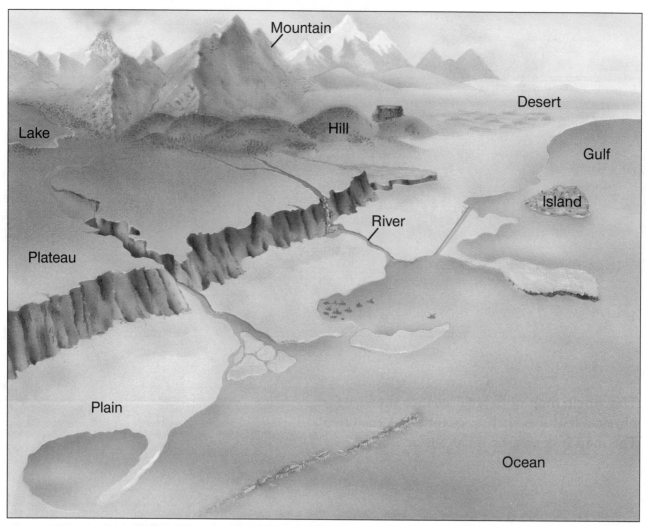

desert – a large, dry area of land with little or no rainfall

gulf – an area of ocean or sea partially surrounded by curved land

hill – a raised piece of land that is smaller than a mountain

Valley –

island – a piece of land surrounded by water on all sides

lake – a body of water smaller than an ocean and surrounded by land

mountain – peaked land that rises very high above the land around it

grassland –

ocean – a huge body of salt water that covers nearly three fourths of Earth's surface

plain – a large area of flat or gently rolling land

plateau – a large, flat area of land that rises higher than the land around it

river – a large stream of water that flows through land

1. Study the map. Read each word and its definition.
   How many kinds of landforms and water forms are shown on the map? _____

2. Find the word **mountain** on the physical map.
   Is a mountain a landform or a water form? If it is a landform,
   write the letter **L** above the word on the map.
   If it is a water form, write the letter **W** above the word on the map.

3. Find the word **river** on the physical map.
   Is a river a landform or a water form? If it is a landform,
   write **L** above the word. If it is a water form, write **W** above the word.

4. Write **L** or **W** above each of the other eight forms of land and water
   shown on the map.

5. Using the information you recorded on the map, list each land and
   water feature in the correct column below.

Land	Water

Lesson 1

ACTIVITY Create a game to learn forms of land and water.

A Geography Game

The BIG Geographic Question

What are the different forms of land and water?

From the article you learned that land and water are the two main physical features of Earth. The map skills lesson showed you some of the different forms of land and water. Now create a game to help you and your classmates learn the names of the different forms of land and water.

A. Remind yourself of the main bodies of land and water on Earth.

1. List the seven continents below.

   a. _____   e. _____

   b. _____   f. _____

   c. _____   g. _____

   d. _____

2. List the four oceans below.

   a. _____   c. _____

   b. _____   d. _____

B. Look back at the different forms of land and water shown on pages 4–5.

1. List the landforms below.

   a. _____   d. _____

   b. _____   e. _____

   c. _____   f. _____

2. List the water forms below.

   a. _____   c. _____

   b. _____   d. _____

C. Write down the names of some of your favorite games. What game could you make to help you learn the names of oceans, continents, landforms, and water forms?

_____   _____

_____   _____

D. Think about how you will teach the game to your classmates. Write simple directions for playing your game on the lines below.

_____

_____

_____

_____

E. Make your game. First, try it out with a family member or friend to see if you need to make any changes. Make the changes, then teach your game to a classmate.

Your Place in Space

As you read about your place in space, think about how you might represent the many places in which you live.

How do you answer when someone asks you where you live? There are many ways to answer this question.

The first thing you might think of is your home. At home, you have a place in the space of your family. You are a brother or sister, a son or daughter. Your family lives in a home. You may think about the rooms in your home. Your home is one of your places in space.

Next you may think about the street where your home is located. That street is part of a community. Maybe a picture of your neighborhood street comes into your mind. Or maybe a picture of you and your family walking in a community park enters your mind. Your community is another one of your places in space.

Look for your community on a city or state map. You may or may not find it there by name, but you will probably find a nearby city that you recognize. Circle or write in your community's name on the map if it isn't already there. The city in which your community is located or near your community is another one of your places in space.

What place do you have in the space of your family?

While looking at the city map, you may notice the state in which the city and your community are located. The state in which you live is another one of your places in space. If you looked at a United States map, you would find your state on it. If you looked at a world map, you would find the United States on it. And if you looked at a map of the solar system, you would find the world, or Earth, on it!

Look at the **mental map,** a way of organizing information in your mind, below. It is one way of thinking about your place in space. Find another example of a mental map in the Almanac in the back of the book.

Yes! Your family, your home, your community, your city, your state, the United States, North America, and Earth are all your "place in space." The picture in your mind of your place in space may have started out small, but it got bigger and bigger! When someone asks you where you live, how will you answer? You'll probably just give your address. But, in your mind, you can picture a lot more.

Home
Community
City
State
United States (Country)
North America (Continent)
World (Earth)

A mental map that represents your place in space, might look like this.

You have a special place in the world.

Lesson 2
MAP SKILLS Making Mental Maps

Mental maps are our ideas of where something is and how to get to it.
They are images we store in our minds. When we can't get our hands
on a real map, mental maps help us remember how to get to what
we want or where we want to go. Sunbury is a small community
located near the bigger city of Columbus, Ohio.
Use this map of Sunbury to practice making
a mental map.

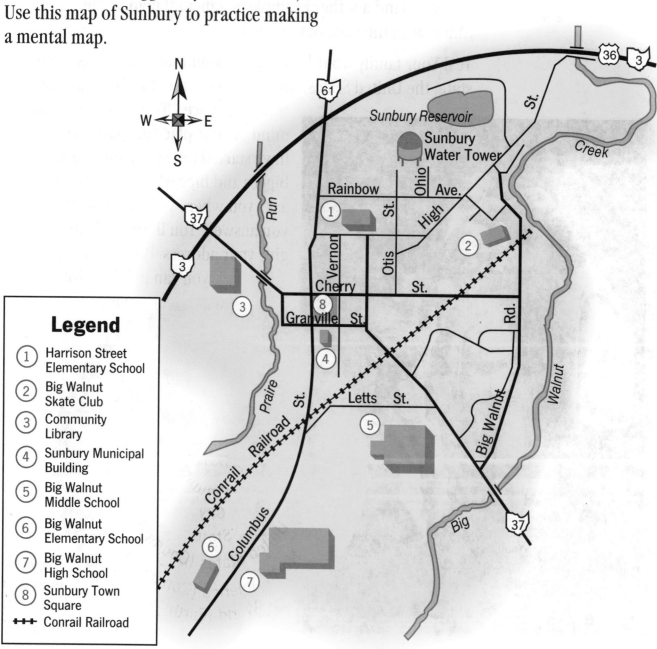

Legend

1. Harrison Street Elementary School
2. Big Walnut Skate Club
3. Community Library
4. Sunbury Municipal Building
5. Big Walnut Middle School
6. Big Walnut Elementary School
7. Big Walnut High School
8. Sunbury Town Square

+++ Conrail Railroad

1. Where is the town square located?

_____

2. Write the names of the four streets that surround the town square.

_____   _____

_____   _____

3. Which two of the four streets listed above run north and south?

_____

4. Which two of the four streets listed above run east and west?

_____

5. List the names of some interesting places in Sunbury and where
they are located.

Place Name	Location

6. Now cover up the map of Sunbury with a piece of paper. Try to
remember what it looked like. Draw a picture of it in your mind.
Using your mental map and your notes from questions 3, 4, and 5
above, draw a map of Sunbury on paper. When you are finished,
uncover the map in your book and compare it to the one you drew.
How accurate was your mental map?

Lesson 2

ACTIVITY
Make a mental map of your own "place in space."

Your Place, Your Space

The **BIG**
Geographic Question

Where do you live?

In the article you learned about the many places that you can say you live. The map skills lesson gave you some practice in making a mental map. Now you will make a mental map of your place in space to show the many places you live.

A. A mental map can be of a big or small space. It can be specific or general.

1. Write the name of your community or town.

_____

2. List some of the land and water features, buildings, and fun places to go in your community.

Land Features	Water Features	Buildings	Fun Places to Go

3. Think about the features you listed in the chart and what your community or town looks like.

4. Draw a mental map of your community or town like you did for Sunbury.

B. Think about the bigger, more general areas that include your community or town. To help you with this, answer the following questions.

1. Near what bigger city is your community or town located?

_____

2. In what state is your community or town located?

_____

3. Use the United States map in the Almanac to locate your state.

4. Use the world map in the Almanac to locate the United States.

C. The "bull's-eye" mental map that you saw on page 9 is one way to show your place in space. Look in the Almanac for an example of a mental map for a person who lives in Atlanta. Now create a mental map that shows your place in the bigger space of the world.

Lesson 3

Making a MAP

As you read about the features or elements of a map, think about a map you'd like to make and share.

A **map** is a flat drawing that represents a place. It shows a bird's-eye view of parts of our world. Some maps are very simple; others are detailed. Either way, maps are important tools for providing information.

When planning a map, a mapmaker must decide two things: what the map will be about and what features are important to show and explain. The most important thing is this— it must be easy to understand!

A **map's scale** helps us figure out how far from each other places on a map really are. The scale may show that one inch on the map equals 10 miles on the ground.

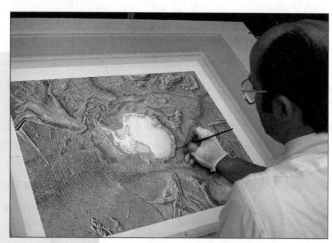

Did you know that the making and study of maps is cartography? Someone who draws maps is called a **cartographer.**

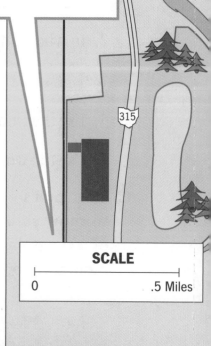

Olentangy River Road

Olentangy River

315

161

315

SCALE

0   .5 Miles

14

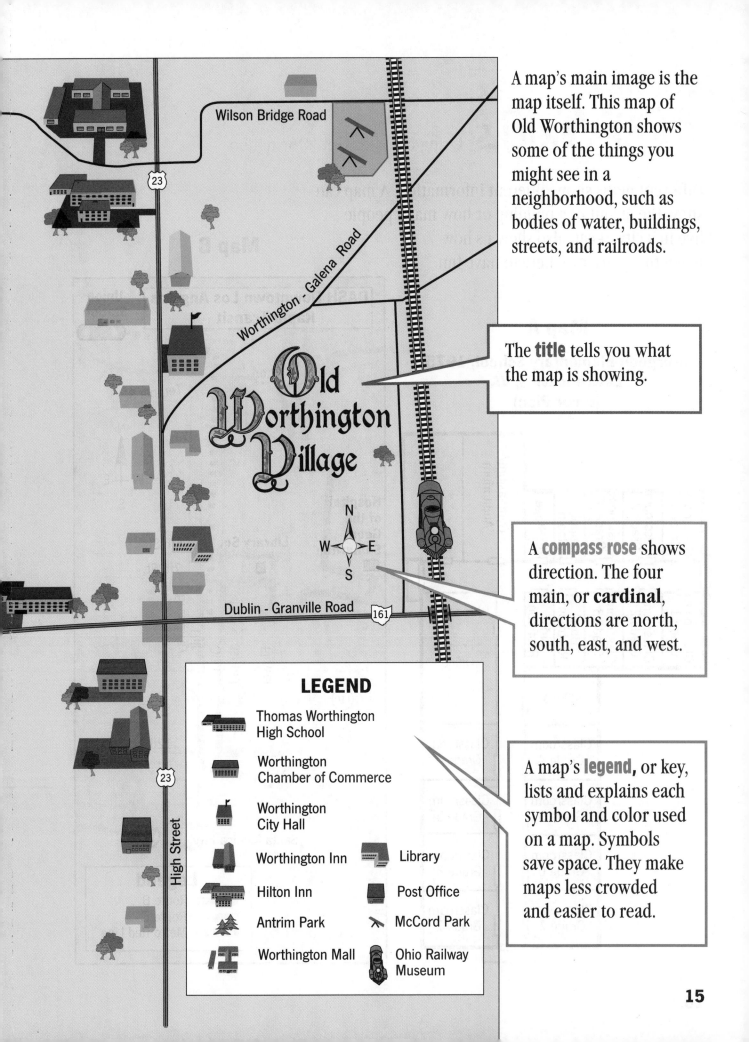

A map's main image is the map itself. This map of Old Worthington shows some of the things you might see in a neighborhood, such as bodies of water, buildings, streets, and railroads.

The **title** tells you what the map is showing.

A **compass rose** shows direction. The four main, or **cardinal**, directions are north, south, east, and west.

A map's **legend**, or key, lists and explains each symbol and color used on a map. Symbols save space. They make maps less crowded and easier to read.

Wilson Bridge Road

Worthington - Galena Road

Old Worthington Village

Dublin - Granville Road

High Street

LEGEND

Thomas Worthington High School

Worthington Chamber of Commerce

Worthington City Hall

Worthington Inn   Library

Hilton Inn   Post Office

Antrim Park   McCord Park

Worthington Mall   Ohio Railway Museum

15

Lesson 3

MAP SKILLS Using Maps to Get Information

Different maps show different information. A map can show the layout of a building or how many people live in a city. It can also show us how to get to a place or where to have fun.

Map A

**Sturgeon Elementary School, 1975
Lawrenceville, VA
(Floor Plan)**

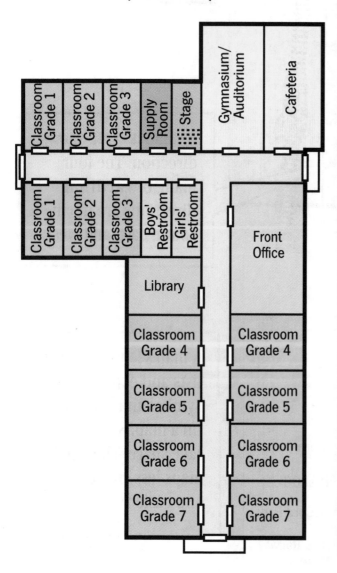

Map B

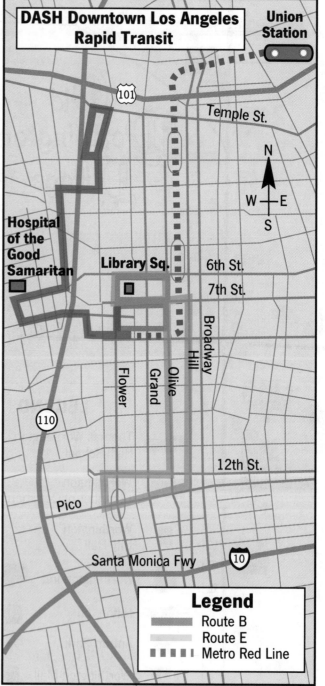

16

A. Look at Maps A and B.

1. Which map tells you which bus to take to Library Square?

_____

2. Which map can help visitors find their way around an elementary school?

_____

B. Answer these questions about Map A.

1. What is the title of Map A? _____

2. Does the school have a gym? _____

3. Draw a line to show how to go from the library to the supply room.

4. Write one sentence describing the information this map gives.

_____

5. What do you think this map can be used for?

_____

C. Answer these questions about Map B.

1. What is the title of Map B? _____

2. Which bus route takes you to Library Square? _____

3. Where does the Metro Red Line take you?

_____

4. Write one sentence describing the information this map gives.

_____

5. What do you think this map can be used for?

_____

_____

Lesson 3

ACTIVITY Make a map that provides information and communicates a message.

Mapping a Message

The **BIG** Geographic Question

How do maps communicate information?

From the article you learned some of the basic features of a map and their purposes. The map skills lesson showed you that different kinds of maps are used to communicate different kinds of information. Now create a map that communicates information to a family member or friend.

A. Read the below list of different types of maps. Circle the type of map that you think you would like to create. You may want to look in the Almanac for examples of other types of maps, such as a recreation or population map. Or you can add your own ideas to the list below.

a map of how to get to your room in your house

a neighborhood street map

a map of your school bus route

a map of the land and water features of a nearby park

a map of where you hid something last year

_____

_____

_____

_____

B. Make a list of things to include on your map. Look back at the article on pages 14 and 15 for suggestions.

_____   _____

_____   _____

_____   _____

C. Think about your map's legend. Choose four things on your map that you could show as symbols. Write one of the names in the space at the top of each box. Then draw the symbols in the larger boxes.

D. Make a practice sketch of your map so you can plan where to put everything.

E. Make a final version of your map. Have a family member, friend, or classmate read your map and describe the information or message shown. Was their answer correct? Did your map communicate the message you wanted? If not, how could you change it to make it more clear?

Lesson 4

People Places

As you read about the town of Petersburg, think about the different people and places that make up a community.

Community services help everyone. How do they help children?

Petersburg is a suburban community near Richmond, the capital city of Virginia. A long time ago, it was an important battle site during the Civil War of 1861–1865. During that time Petersburg was mostly farmland where tobacco was grown. Now there are more houses than fields in Petersburg, but farming is still important to the community.

Today Petersburg is a spread-out surburban community. A lot of people there live in subdivisions, or areas of land divided into lots for building houses. Many of the subdivisions have recycling programs. This means that the community members are finding ways to reuse things instead of throwing them away.

The Petersburg community has post offices, libraries, hospitals, police and fire stations, federal government offices, and other services. There are several schools in Petersburg. The town has also just built a shopping mall near the interstate highway. This location makes the mall easy to get to by people who live in Petersburg and people traveling through the area.

Many of the people who live in Petersburg have jobs downtown. They work in stores and offices. Other people who live in Petersburg go to jobs in the city of Richmond, less than 30 minutes away. Some of them take the bus to work every morning, and some of them drive.

On the edge of town is the Appomattox River. Long ago plantation owners used the river to travel and to send and receive goods. Today businesses use the river to ship goods to and receive goods from other cities. The people of Petersburg enjoy boating and fishing on the river.

The people of Petersburg work together and play together. They depend on each other for goods and services and to have fun. They know that a community is a place, but it is people, too.

These boys and girls are on their way to school. Do you ride a bus to school?

MAP SKILLS Using a Map Compass

If you were a visitor in Petersburg, a map could help you find your way around. Many maps have a symbol called a compass or compass rose that shows the four cardinal directions—north, south, east, and west.

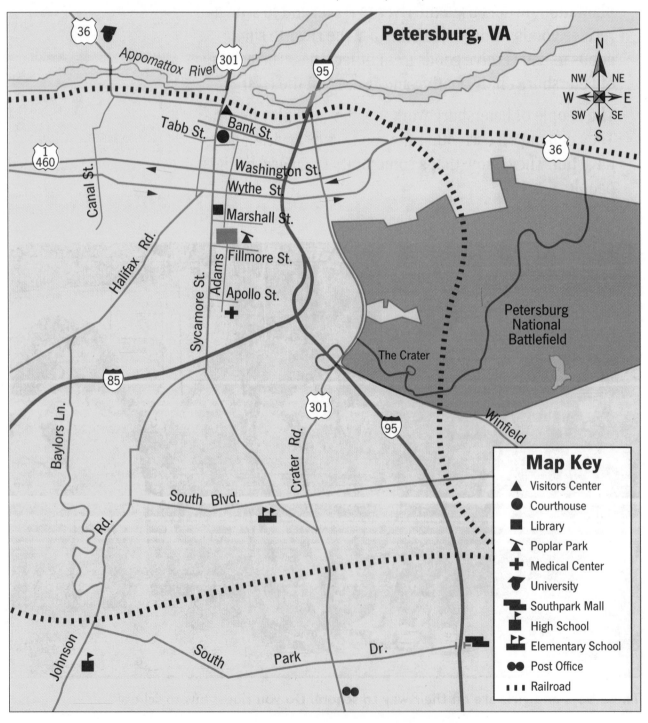

Petersburg, VA

Map Key

▲ Visitors Center
● Courthouse
■ Library
⚑ Poplar Park
✚ Medical Center
⛊ University
🏬 Southpark Mall
🏫 High School
🏫 Elementary School
●● Post Office
▪▪▪ Railroad

A. If you were at Petersburg National Battlefield, which direction would you go to get to these places?

1. medical center _____

2. post office _____

3. visitors center _____

Northeast is an **intermediate direction**—a direction halfway between two cardinal directions. The other intermediate directions are northwest, southeast, and southwest.

4. In which direction would you travel from the battlefield to the

   Appomattox River? _____

5. In which direction would you go from the Medical Center to

   Southpark Mall? _____

B. Follow these directions. Where are you?

1. Start at the high school. Go north on Johnson Rd. to Sycamore St. Go south on Sycamore St. to South Blvd.

   _____

2. Start at the post office. Go north on Crater Rd. to Washington St. Go west on Washington St. to Sycamore St. Go south on Sycamore St. Stop at the corner of Marshall St.

   _____

C. Now that you can find your way around Petersburg, you can give directions to someone else.

1. Tell how to get from the high school to Poplar Park.

   _____

2. Tell how to get from the Courthouse to Southpark Mall.

   _____

Lesson 4

ACTIVITY Use a map to show how to find a place in your community.

Where Are You?

The **BIG** Geographic Question

How can you help visitors find their way around your community?

In the article you read about some of the people and places in the community of Petersburg. The map skills lesson helped you find your way to some of the important places. Now use what you know about directions to make a map of your community.

A. Make a list of important places in your community. An example has been done for you.

   1. school

   2. _____

   3. _____

   4. _____

B. Think of symbols to stand for the places on your map. Draw a symbol in each box.

1.

2.

3. [box]

4. [box]

C. List at least five streets that will be on your map.

1. _____

2. _____

3. _____

4. _____

5. _____

D. Draw a map of your community. Be sure your map includes all the places and streets from your lists. Also include a map key.

E. Write directions.

1. Choose one of the places on your map. Draw the way to go from your school to that place.

2. Write directions from your school to that place. Be sure to include street names and compass directions.

_____

_____

_____

It's a PLAN!

As you read about the plan for Washington, D.C., think about how geography affected where the city is and how it was designed.

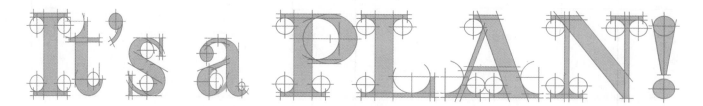

L'Enfant's plan made the Capitol building the center of the city.

The year was 1790. The United States did not have a permanent capital. Instead, the government had moved eight times from one city to another. Finally the nation's leaders decided the government should have a permanent home. But where?

President George Washington was asked to choose the site for the nation's new capital. In 1791 he picked a location between the states of Maryland and Virginia. The decision to place the capital on the Potomac River was based on geography. Washington believed that the location would provide a good port. The new capital would also be centrally located in the nation.

Pierre-Charles L'Enfant, a young French engineer, was chosen to plan the capital. L'Enfant believed in the United States' future. He had grand ideas for the new capital.

In fact, some of L'Enfant's ideas were too grand. But he would not compromise, or give up any of his plans in order to reach an agreement. In 1792 President Washington dismissed L'Enfant and found someone else to plan the capital city.

Andrew Ellicott and Benjamin Banneker were the new planners. Using parts of L'Enfant's original plan, they put the Capitol, the building where Congress meets, in the center of the city.

Ellicott and Banneker laid out the streets in a **grid** pattern to make the city easy to travel across. A grid is a system of lines that forms boxes. The streets that ran east and west were named with letters. The streets that ran north and south were named with numbers. Diagonal streets were named for states of the union. Many streets met in the middle of the city in hubs, or circles.

In 1800 the government moved to its new home. A well-organized street plan and a central location near a major waterway made it a true "capital" city!

L'Enfant's plans included the Mall, a broad, green park lined with many important buildings. One of them is the Washington Monument.

Lesson 5
MAP SKILLS Using a Grid to Locate Places

Using a map grid can help you find places on a map easily. Look at the map of Washington, D.C., below. Suppose you want to find the Capitol building. First, you look for the word *Capitol* in the Map Index. Next to *Capitol* in the index, it says E17.

To find grid box E17, find the letter *E* along one side of the map. Then move your other finger across the top or bottom of the map until it is under the number *17*. The box where *E* meets *17* is where you will find the Capitol building.

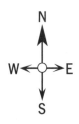

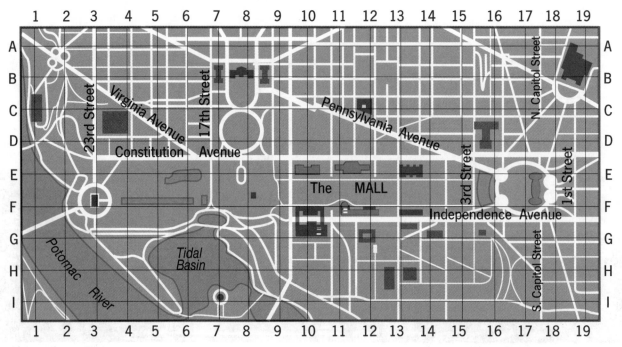

Map Index

Capitol	E 17	National Air and Space Museum	F 13
FBI Building	C 12	National Gallery of Art	E 13
Jefferson Memorial	I 7	Smithsonian Institution	F 11
Lincoln Memorial	F 3	Union Station	A 19
Museum of American History	E 10	Washington Monument	F 8
Museum of Natural History	E 11	White House	B 8

A. Use the map index and grid to find these places.

1. Lincoln Memorial _____

2. FBI Building _____

3. Smithsonian Institution _____

B. Use the map grid to identify these places.

1. What important home is found at B8? _____

2. A memorial to the third U.S. president is found at I7. What is its name?

_____

3. A major transportation center is located at A19. What is it? _____

_____

C. Use the map index and grid to answer these questions about the Mall.

1. What is at the east end of the Mall? _____

2. What is at the west end? _____

3. Identify four of the important buildings found along the Mall.

_____

_____

_____

_____

Lesson 5

ACTIVITY
Compare your community's history with the history of Washington, D.C.

Your Community: Was It Planned?

The BIG Geographic Question

How can you tell whether you live in a planned community?

From the article you learned how the community of Washington, D.C., was planned. The map skills lesson showed you how to use a grid system. Now find out about the history of your own community to figure out whether it was planned.

A. Answer these questions to find out about the history of your community.

1. Who were the first people to settle in your community? _____

_____

_____

2. When was the community started? _____

_____

_____

3. Why was the community started?_____

_____

_____

B. Complete the chart below to compare facts about your community to Washington, D.C.

	History of Washington, D.C.	History of Your Community
Why was that location selected?		
What were its major physical features?		
What is the purpose of the community?		
Were streets planned using a grid?		
Were parks included in the plan?		
Where are many important buildings located?		

C. Think about how your community grew and complete the following.

1. Find or make a map of your community as it is today.

2. Tell how your community has grown.

_____

_____

_____

3. Do you think your community was planned?

_____

_____

_____

_____

Four Fabulous Communities

As you read about these communities, think about the geographical features that make them unique.

Wind has shaped the Arizona plateau.

Over 800 years ago the Hopi Native American tribe settled in the town of Oraibi, Arizona. This community is in the sandy red **plateaus**—or high, flat lands—of northern Arizona. There, wind and rivers have carved the land into amazing shapes, cutting steep canyons out of the red rock.

The Hopi make beautiful pottery. Its patterns and colors reflect the canyon walls and the colors of the rocks of the Arizona plateau.

Another state with a unique Native American community is Oklahoma. It has the largest Native American population in the United States. Each June the Red Earth Native American Cultural Festival is held in Oklahoma City. People from all over North America go to the festival.

More than fifty tornadoes roar through the state of Oklahoma each year.

Oklahoma is in the **plains.** This area of broad, flat land gets little rain, but there is lots of wind. No wonder the state song says "Oklahoma, where the wind comes sweepin' o'er the plain."

Water, not wind, makes Leech Lake in northern Minnesota special. For more than 400 years the Ojibway have gathered wild rice from the lake. They harvest the rice from canoes, knocking the grains into their boats as they glide through the quiet waters.

Northern Minnesota is full of rivers and lakes. Dense forests protect many kinds of wildlife. It's one of the few areas in the country where timber wolves still live.

Minnesota is known for its beautiful lakes.

Forests also cover the hills of western Virginia, where the town of Abingdon is located. The hills aren't high, but they're rugged. Long ago, it was hard to travel to get goods for the home. For this reason, families made most of their household items. Travel is easier now, but many residents still make traditional crafts.

These are just four examples of how physical features such as plateaus, plains, hills, rivers, and lakes can help make a place unique. In many communities in the United States, people use the physical features of their area to create items or events that they can share with people who do not live there.

The hills of western Virginia are rugged.

MAP SKILLS Using a Map to Locate Capital City Communities

Every state has many unique communities. But every state has only one capital city. A state's capital is the home of its government. It is where the state governor holds office and where the state legislature meets. These are some of the things that make a capital city unique. But, is the capital city representative of other unique communities in a state?

Richmond

Abingdon

Virginia Beach

Leech Lake

St. Paul

Minneapolis

WA

OR

ID

MT

ND

WY

SD

NV

UT

CA

CO

NE

KS

AZ

NM

TX

OK

MN

WI

IA

IL

MO

AR

MS

LA

MI

IN

KY

TN

AL

OH

WV

VA

NC

GA

SC

FL

PA

NY

ME

VT

NH

MA

CT

RI

NJ

DE

MD

Washington, D.C.

Scale

0   400 km

0   400 Miles

AK

HI

Oraibi

Phoenix

Tucson

Stillwater

Tulsa

Oklahoma City

Map Key

⊛ United States Capital City

★ State Capital City

● City

A. Different maps use different symbols to show the capital city of a state or country. You can find each symbol in the map key. Look at the map of the United States. Find the map key.

   1. What is the symbol for the United States capital? _____

   2. What is the symbol for the state capitals? _____

B. Find the states of Arizona, Minnesota, Oklahoma, and Virginia on the map. Use the information on the map to find each state's capital and the community described in the article for each state. Record your information on the chart below.

State	Capital City	Community Featured in Article
Arizona		
Minnesota		
Oklahoma		
Virginia		

C. Choose one of the capital cities you listed above to learn more about. Look in the Almanac for information on the state's capital. Complete the T-chart below by listing information about the capital city and the featured city. Note any similarities and differences. Based on the information you have collected, decide whether the state's capital city is representative of other unique communities in the state.

Capital City	Featured City

Lesson 6

ACTIVITY

Advertise physical features of your community that attract people to live or visit there.

Your Community: What Are Its Features?

The **BIG** Geographic Question

What physical features of a community make it a unique place?

In the article you read about some of the physical features that make four communities special. In the map skills lesson you located state capitals and looked at whether they were representative of other communities in the state. Now plan and present an advertisement that highlights a physical feature of your community to attract people to come there.

A. Think about the features of the communities you read about on pages 32–33.

1. Use the chart below to list the important physical features of each community. An example has been done for you.

State	Arizona	Minnesota	Oklahoma	Virginia
Community	Oraibi			
Features	• sandy red plateaus • wind and rivers carve canyons through red rock			

2. How do these features help make the communities special places to live or visit?

_____

_____

_____

B. Think about your community and complete the following.

1. Make a chart showing the special physical features of your own state, its capital city, or your community. Some of these physical features might include rivers, lakes, forests, mountains, or plains. Some other features might include the different groups of people who live there or fun things to do there. Jot down your notes below.

Your Community's Name	
Special Physical Features	
Other Special Features	

2. How do these features make your community special?

_____

_____

_____

C. Now plan and present an advertisement for your community. Your advertisement can be a poster, a flyer, a TV commercial, or a radio commercial. It might include a picture and a written description of the physical feature that makes your community a unique place to live or visit.

Lesson 7

Adventure to the New World

As you read about Jamestown, think about the role geography played in its settlement.

Jamestown has been restored to the way archaeologists believe the original village looked.

Would you dare to leave everything behind and travel to an unknown place far away? How would you know where to settle? What would you look for to help you survive? In December of 1606 three small wooden ships left London, England, heading for the New World. The people on board traveled through good and bad weather. They were hoping for a better life in the New World. Many believed they would find silver or gold and get rich quickly.

The journey was long and hard. After more than four months, the colonists finally spotted land. Did they anchor their ships near the first place they saw? No. They followed the James River until they came upon a little peninsula. A **peninsula** is an area of land surrounded by water on three sides. The colonists thought this would be a safe location, since it would be hard for attackers to see from the ocean. It would also be possible to defend from both land and sea. In addition, there were many trees to use for building the things necessary to survive.

The colonists settled on the peninsula and named it Jamestown. This settlement was named after the English king, James I.

Was this location a good choice? Sadly, it was not. The ground was swampy. The only drinking water was from the river, which was full of germs. Many people became sick and died. Diseases, starvation, and attacks by Native Americans also took the lives of many colonists.

In spite of the problems, Jamestown did survive and was the first permanent English settlement in North America. It became an important Virginia farming community, as well as a model for early government in America.

The peninsula that was the site of the original Jamestown settlement is now an **island.** An island is an area of land surrounded by water on all sides. Much of the original land has been washed away by currents in the James River. Luckily, however, the Jamestown settlement has been restored to the way it looked in the 1600s. Each year thousands of people visit Jamestown, Virginia, to see, taste, touch, and hear what life was like when our country was just beginning.

VIRGINIA'S
Discovery of
SILKE-VVORMES,
with their benefit.

AND
The Implanting of MULBERRY TREES.

Also
The dressing and keeping of Vines, for the rich Trade of making Wines there.

Together with
The making of the Saw-mill, very usefull in *Virginia*, for cutting of Timber and Clapbord, to build withall, and its conversion to other as profitable Uses.

LONDON,
Printed by *T. H.* for *John Stephenson*, at the Signe of the Sun, below Ludgate. 1650.

The harbor of Jamestown, Virginia, as seen from an airplane.

MAP SKILLS
Using a Historical Map to Get Information About the Past

A **historical map** gives information about the past. It can give information about the physical and human features that caused things to happen in the past.

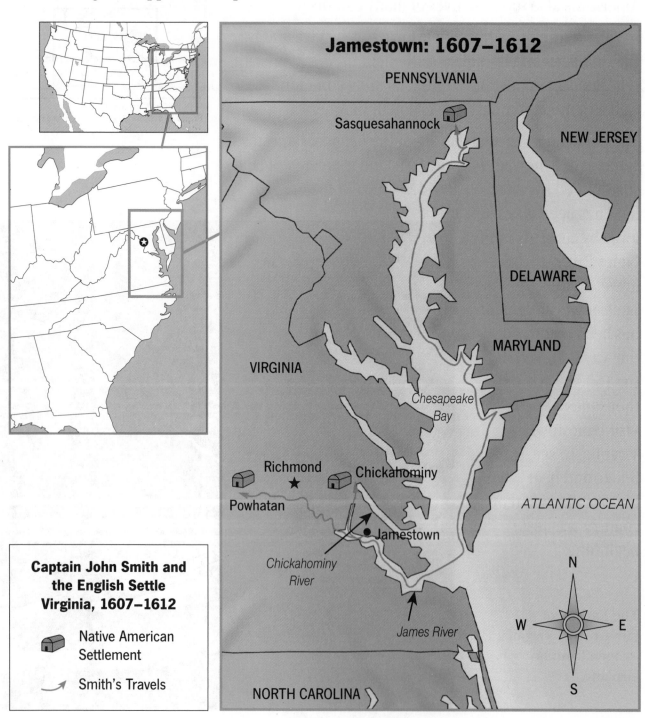

Jamestown: 1607–1612

PENNSYLVANIA

Sasquesahannock

NEW JERSEY

DELAWARE

MARYLAND

VIRGINIA

Chesapeake Bay

Richmond   Chickahominy

Powhatan

Chickahominy River

● Jamestown

ATLANTIC OCEAN

James River

N
W   E
S

NORTH CAROLINA

Captain John Smith and the English Settle Virginia, 1607–1612

🏠 Native American Settlement

↗ Smith's Travels

A. Look at the map of early Virginia.
1. Circle Jamestown.
2. Write down the time in history that this map is showing._____
3. Circle the Chesapeake Bay and Atlantic Ocean.
4. Mark an *X* on the early Native American settlements.

B. Look at the map and complete the following.
1. Circle the names of the bodies of water you see on the map.
2. Write the names of the bodies of water.

    a. _____   c._____

    b. _____   d._____

3. Trace the peninsula on which Jamestown is located.
4. Write the names of the bodies of water Captain Smith followed to get from Jamestown to:

    a. Susquesahannock (səhs-kwəh-səh-HAN-ək)

    _____

    b. Powhatan (pow-HAT-əhn)

    _____

    c. Chickahominy (chi-kə-HÄ-mə-nē)

    _____

5. Was the peninsula a good location to settle Jamestown? Why or why not?

    _____

    _____

6. Why was it important to be located near water in the early days of the Jamestown settlement?

    _____

Lesson 7

ACTIVITY
Decide whether you would have left or stayed in Jamestown.

What Would YOU Do?

The BIG Geographic Question

Would you have remained in Jamestown or gone back to England in 1607?

A. List on the idea web below characteristics of life in early Jamestown. Include both positive and negative characteristics. Add more spaces if you need to.

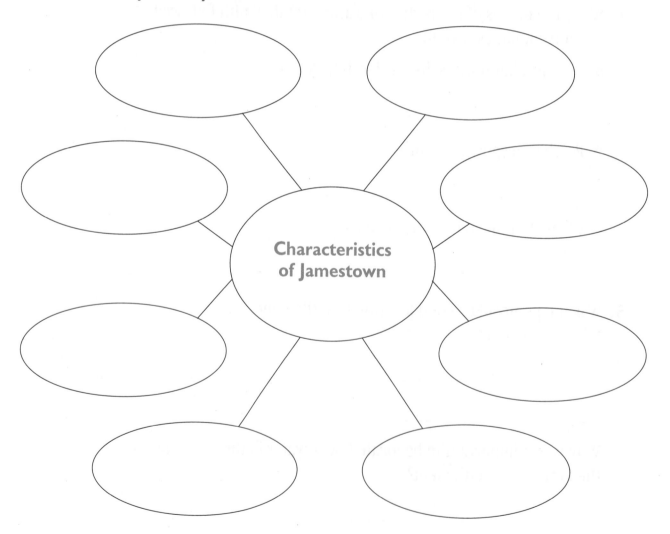

Characteristics of Jamestown

B. Use the chart below to take notes on what the living conditions were like in Jamestown in 1607.

	Good	Bad
Land Features		
Water Features		
Other Features		

C. Using the information you collected, list reasons you would leave or stay in Jamestown.

Reasons to Leave	Reasons to Stay

D. Pretend to be an early Jamestown settler. With your classmates, stage a town meeting to discuss whether to leave or stay in the community. Listen carefully to the ideas of others. When the meeting is over, think about the arguments you heard both for staying and for leaving. What did you decide? Write your decision and the reasons for it on the lines below.

_____

_____

_____

_____

Our First Communities

As you read about the first North Americans, think about how they used the land to live.

Thousands of years before the first Europeans arrived in North America, the first Americans traveled across the land and settled. These Native Americans of long ago **adapted** to, or learned to live in, the natural environments. Some groups hunted animals and gathered food. Others grew crops. From the land and its resources, they made many of the everyday objects they needed.

The Iroquois of long ago made colorful wampum belts from shells and beads. They wove baskets and made masks from cornhusks. From the trunks of large trees, they carved masks and wooden bowls. The Iroquois used bark from the oak, birch, and elm trees of the northeast woodlands to make canoes. The homes of the Iroquois were made of wooden poles covered with bark and were filled with objects they made from the land.

Square Tower House, Mesa Verde Ruins. No one knows why the Anasazi, an ancient people, moved from the cliff dwellings 700 years ago.

The Sioux Native American tribe of long ago made their home in the Great Plains. Many of the objects they made came from the animals that lived on the plains. These Native Americans carved cups, bowls, and tools from animal bones. The Sioux made leggings, shirts, and moccasins from deerskin. They used buffalo hides draped over wooden poles to form tepees. These Native Americans of long ago found a use for almost every part of the buffalo to make everyday objects like drums, pots, and balls for play. They even used buffalo hooves to make rattles and sewed with buffalo sinew for thread and bones for needles.

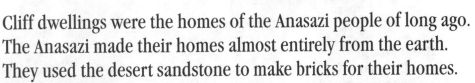

Cliff dwellings were the homes of the Anasazi people of long ago. The Anasazi made their homes almost entirely from the earth. They used the desert sandstone to make bricks for their homes.

These bricks were held together with **adobe,** a sandy clay mixed with straw. The Anasazi discovered that adobe was easy to mold and would not shrink in the hot, dry weather. These Native Americans of the southwest also made beautiful pots from clay. They often painted elaborate geometric designs on them.

Today we can learn much about the very first North Americans by studying the **artifacts,** or tools, shelters, crafts, clothing, and other objects, they made. Artifacts from the past provide us with the clues we need to put together a picture of life in North America long ago. They are clues to how Native Americans, like other early Americans, adapted to their environment.

Lesson 8
MAP SKILLS Using a Map to Show Land and People

Some map keys use color to show the landforms and water forms or to show where groups of people live. Other map keys use symbols to show the same information.

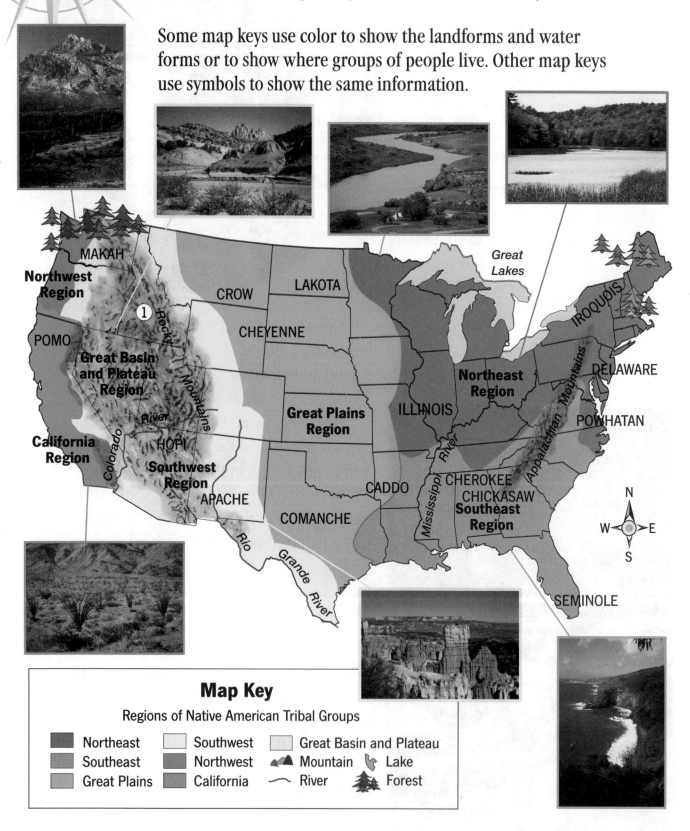

MAKAH

Northwest Region

CROW

LAKOTA

Great Lakes

CHEYENNE

IROQUOIS

POMO

Great Basin and Plateau Region

DELAWARE

Northeast Region

Rocky

Mountains

River

ILLINOIS

POWHATAN

Appalachian Mountains

California Region

HOPI

Southwest Region

APACHE

Colorado

CADDO

Mississippi River

CHEROKEE

CHICKASAW

Southeast Region

Great Plains Region

COMANCHE

Rio

Grande River

N

W   E

S

SEMINOLE

Map Key

Regions of Native American Tribal Groups

Northeast   Southwest   Great Basin and Plateau

Southeast   Northwest   Mountain   Lake

Great Plains   California   River   Forest

46

A. Look at the map key. Put the appropriate number on the map to show where each of the following physical features is located. An example has been done for you.

1. mountain   **3.** lake   **5.** Great Plains

2. river   **4.** forest   **6.** ocean

B. Look at the map and map key to regional groups.

1. Which color shows where the Iroquois tribe lived? _____

2. The southwestern United States has many desert areas. Which Native Americans of long ago lived in this area? _____

3. Which groups of Native Americans might have used resources from the ocean? _____

4. On what river might Native Americans from the Great Plains region have traveled north and south? _____

5. Which groups obtained their resources from forests?

_____

6. What three regions were divided by mountains?

_____

7. How do you know that the Chickasaw and the Cherokee came from the southeast?

_____

8. If you traveled by land from your state to visit a Native American tribe of long ago in the northwest, what landforms or water forms would you have had to cross? If you lived in the northwest, think about traveling to visit a tribe in the southeast. Explain whether your trip would be difficult.

_____

_____

Lesson 8

Find out about the ways of living of Native American tribes long ago.

Digging for Clues

 The **BIG** Geographic Question

What can you find out about the first Native Americans by studying the artifacts they left behind?

From the article you learned about the tools, crafts, clothing, and shelters of some Native American tribes of long ago. The map skills lesson showed you where they lived and some of the landforms found in those regions. Use these clues and information you find in the Almanac to learn about the way of living of a Native American tribe of long ago.

A. Complete the following.

1. List the names of Native American tribes you have read about.

_____   _____

_____   _____

_____   _____

_____

2. Write down the name of a Native American tribe that you would like to learn more about.

_____

_____

Rural communities are very important ones. The families in these communities take pride in the fruits, flowers, grains, and vegetables they grow and the animals they raise. Often these communities have fairs and festivals to show and celebrate the products that are important to them as well as us. The products that rural communities raise and the fairs and festivals they have are an important way that people interact with the geography around them.

CORN PALACE FESTIVAL
Mitchell, SD
Sept. 8–17

VERMONT APPLE FESTIVAL
Springfield, VT
Oct. 7–8

CRAWFISH FESTIVAL
Breaux Bridge, LA
May 7–8

WEST VIRGINIA PUMPKIN FESTIVAL
Milton, WV
Oct. 6–8

MT
ND
MN
ME
VT
NH
SD
WY
NY
MA
CT
RI
IA
WI
MI
PA
NJ
NE
OH
DE
MD
UT
CO
IL
IN
Washington, D.C.
WV
VA
KS
MO
KY
NC
OK
TN
AR
SC
NM
MS
AL
GA
TX
LA
FL

MAP SKILLS
Using a Map Key to Find Out Where Products Come From

A map key explains the symbols used on a map. Sometimes symbols look like the things they stand for. A map key can help us understand a map.

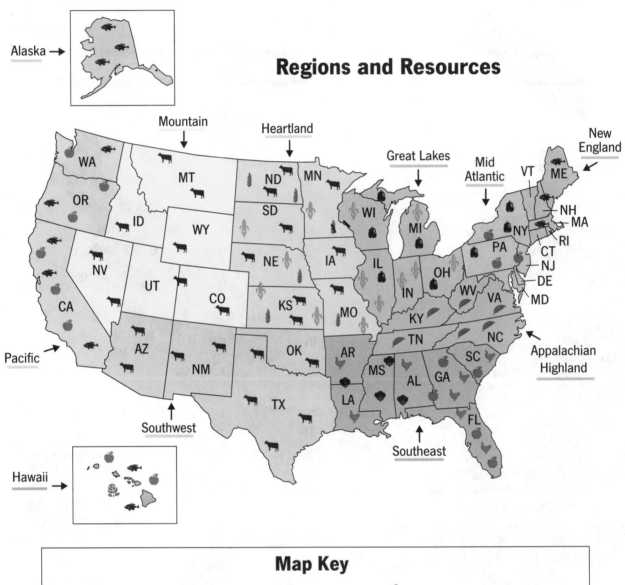

Regions and Resources

Alaska →

Mountain
Heartland
Great Lakes
Mid Atlantic
New England

WA
MT
ND
MN
VT
ME
OR
SD
WI
NH
MA
ID
WY
MI
NY
RI
NV
NE
IA
IL
PA
CT
UT
CO
IN
OH
WV
VA
NJ
DE
MD
CA
KS
MO
KY
AZ
OK
AR
TN
NC
Pacific →
NM
MS
AL
GA
SC
Appalachian Highland
Southwest
TX
LA
FL

Hawaii →

Southeast

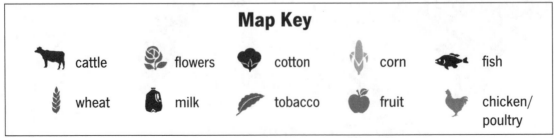

Map Key

cattle	flowers	cotton	corn	fish
wheat	milk	tobacco	fruit	chicken/poultry

A. Cover up the map key on page 58. Now look at the symbols below. Write what each symbol stands for.

(cow) _____

(wheat) _____

(corn) _____

(apple) _____

(fish) _____

(rose) _____

(milk jug) _____

(chicken) _____

(leaf) _____

(cotton) _____

B. Write the name of a product that comes from each region listed below. Then write the states where each product is found.

Region	Product	State
New England		
Mid-Atlantic		
Southeast		
Heartland		
Southwest		
Pacific		
Alaska		
Hawaii		

C. List the regions that produce the most:

cattle _____

corn _____

tobacco _____

Lesson 10

ACTIVITY

Plan a fair to show the products a rural community might be known for and proud of.

Come to the Fair

The **BIG**
Geographic Question

How is a fair or festival in a rural community an example of people interacting with the environment?

From the article you learned of some rural communities that hold fairs and festivals. The map skills lesson showed you products people grow on farmlands or catch in nearby waters. Now plan a fair to display a product of which a rural community is proud.

A. **Use information from the article, the map, and the Almanac to help you choose a place and a product.**

1. Choose a small town. _____

2. Describe the location of the town. _____

   _____

3. Write the name of a product that is important to the town.

   _____

4. Tell whether the product grows on the land or comes from the water.

   _____

B. **Explain why you chose the place and product.**

   _____

   _____

   _____

C. **Read the events on the checklist below. Decide which ones you will have at your fair and write *yes* or *no* in that column on the chart. Describe the events you will have at the fair. Remember to feature the local product.**

Event	Yes or No	Description
parade		
displays		
floats		
food		
games		
posters		
music		
contests		
craft booths		
sales		
speakers		
the mayor		

D. **Make a poster to advertise your festival. Include these details.**

- festival name
- place
- date
- events
- artwork

Lesson 11

A City Comes to Life

As you read about the urban community of St. Louis, think about what makes up a city.

St. Louis, Missouri, is in the heart of the United States. It was once mostly **prairie**—fertile land where grass grew eight feet tall. How and why did St. Louis change from prairie land, to a small town, to a major city?

Like many major cities, St. Louis grew into a bustling city partly because of its location near major waterways. Settlers, traders, explorers, trappers, and gold seekers of long ago found the area easy to reach. St. Louis began as a trading center and grew into a city.

Major cities that are located on important waterways often become big manufacturing centers. Many people move from rural areas to the cities because they can find jobs there. Then shops, stores, restaurants, schools, and hospitals spring up. As more people come to find work, the city keeps growing.

Being located near water and having a reliable system of transportation play key roles in which places become cities. People and culture, jobs and recreational activities, and business and trade are also very important things that make a city grow.

Scientists and engineers work in high-technology industries.

Two mighty rivers, the Mississippi and the Missouri, meet just north of St. Louis. Eight bridges span the Mississippi River which is the site of the St. Louis Gateway Arch.

MAP SKILLS
Using a Map and Chart to
Find Out Where People Live

Mountains, lakes, and rivers are some of the physical features that
can be represented on a map. Population is one of the human features.

**A. Look at this map of the United States. Write the names of the
following physical features that are shown.**

1. Rivers _____

2. Mountains _____

3. Lakes _____

4. Oceans _____

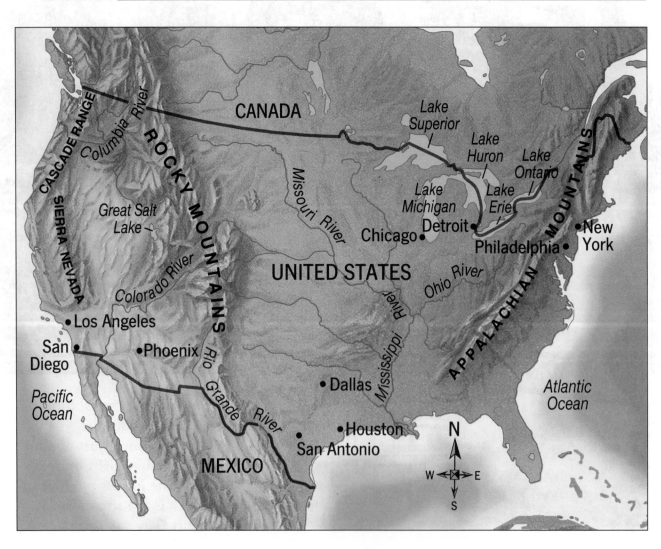

B. Read the city names and their populations on the chart.

1. Complete the chart by ranking the cities from 1 to 10 from largest to smallest population.
2. Locate each city on the map.

Population of Large Cities in the United States

City	Population*	Rank
Chicago, IL	2,768,483	
Dallas, TX	1,022,497	
Detroit, MI	1,012,110	
Houston, TX	1,690,180	
Los Angeles, CA	3,489,779	
New York, NY	7,311,966	
Philadelphia, PA	1,552,572	
Phoenix, AZ	1,012,230	
San Antonio, TX	966,437	
San Diego, CA	1,148,851	

*Population as of July 1, 1992

C. List the cities on the map that are located near each of the following:

Rivers _____

Lakes _____

Oceans _____

D. Why do you think so many large cities are located on or near water?

_____

65

Lesson 11

ACTIVITY Explore an important feature of an urban community.

Investigating a City

The **BIG**
Geographic Question

What makes an urban community a city?

From the article you learned how one city grew to become a big urban area with increased business, transportation, and population. In the map skills lesson you looked at the United States' largest cities that are located near water. Now investigate the human features of your own or a nearby urban community.

A. Listed here are five important features of a city. Describe what each one means. Use the Glossary or Almanac in the back of your book to help you.

1. Population _____

_____

2. Transportation _____

_____

3. Business _____

_____

4. Manufacturing _____

_____

5. Cultural activities _____

_____

B. Complete the following to find out more about important features of an urban community.

1. Write the name of the urban community where you live or the one

   closest to you. _____

2. Estimate the population of the city. _____

3. Fill in this chart with examples you think describe the city you chose.

Features	Examples
Population	
Transportation	
Business	
Manufacturing	
Cultural Activities	

4. Check with a family member, librarian, or encyclopedia to see whether the examples you listed above are correct.

C. Choose one of the features of the city you selected to explore further. Jot down some notes about the feature.

_____

_____

_____

D. Design a banner that celebrates and tells something about the community you studied. The banner design should emphasize the feature on which you focused.

Lesson 12

The 'Burbs

As you read about the suburbs, think about how they came to be and why people choose to live there.

When the United States was young, settlers who wanted more land went west. Today when people want more space, they leave the city and move to the **suburbs.** A suburb is a community of homes located at the edge of a city. Almost half of all people in the United States now live in suburbs.

In the 1880s streetcars, and later commuter trains, improved travel out of cities. People could live in towns or villages and travel by trolley or train to and from city jobs.

Since the 1940s roads and highways have become bigger and better. Real estate developers have built large clusters of homes on what was once open land or farmland. Many families have moved into these homes in the suburbs looking for quiet, friendly communities where people know and look out for each other.

Many villages and towns have become the suburbs of a large nearby city.

68

Families are not the only people moving to the suburbs. Companies locate outside the city, too. Many companies don't want to depend on local materials or markets. They locate near highways, railroad centers, and major airports so they can transport supplies and products to faraway places by truck, train, and plane.

Today more and more people live in one suburb and **commute,** or travel, to another suburb to work. Rarely do these people go to the nearby big city that caused the suburb to grow. When they do, it is usually for entertainment.

Some suburban communities are large tracts of homes that cover big stretches of land away from the city.

Living in the suburbs often means spending mornings and evenings on the highways commuting to and from work in the city.

Interesting Things to Note

- All cities began as "walking" cities and remained as such until new ways of transportation developed in the late nineteenth century.

- People from farms or small towns move to the suburbs, too. These people usually move to the suburbs to have better access to places like grocery stores, hospitals, and schools.

- Some companies allow workers to "telecommute." These workers avoid traffic by working at home using computers, fax machines, and phone lines to connect to the office.

Large interstate highways connect outer suburbs to the city.

69

Lesson 12
MAP SKILLS Using a Map to Understand Geographical Terms

A **metropolitan area** is a large city and its inner and outer suburbs.
The city must have a population of 50,000 or more. The inner suburbs
are the residential areas closest to a city. The outer suburbs are
residential areas that develop as highways extend farther out.

A. Look at the metropolitan area map of Phoenix.

1. Color the city red.

3. Color the outer suburbs green.

2. Color the inner suburbs blue.

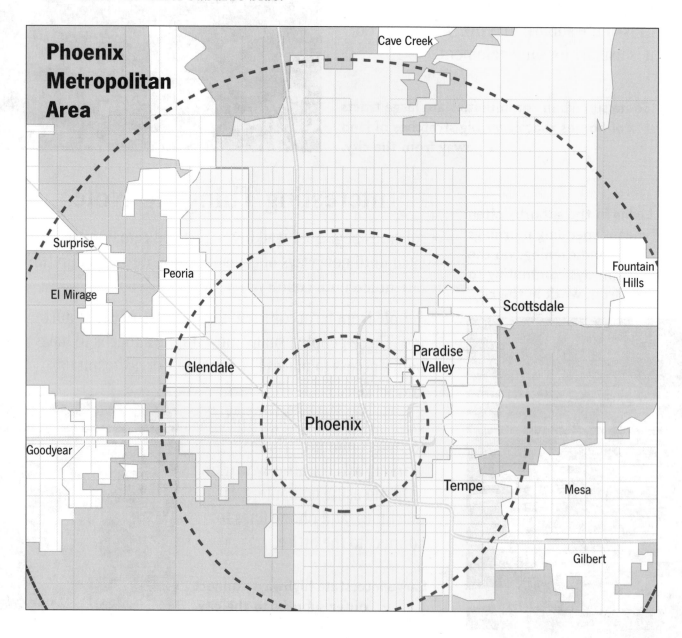

Phoenix Metropolitan Area

Cave Creek

Surprise

Peoria

Fountain Hills

El Mirage

Scottsdale

Glendale

Paradise Valley

Goodyear

Phoenix

Tempe

Mesa

Gilbert

B. **Complete the following about the area surrounding where you live.**

    1. Write the name of the town in which you live. _____

    2. Write the names of other suburban areas near you. _____

    _____

    3. Write the name of the largest city near you. _____

C. **Using the map of Phoenix as a model, draw the area in which you live. Include your town, other cities around your town, and the nearest large city.**

D. **On the map you drew, circle and label the city, inner suburbs, and outer suburbs of the area in which you live.**

Lesson 12

ACTIVITY Explore how a nearby suburban community developed.

Exploring a Suburb

> The **BIG** Geographic Question
>
> **As a suburb develops, how does the area change?**

From the article you learned that people leave both the city and the country to live in the suburbs. The map skills lesson helped you look at the area in which you live in relation to suburbs. Now find out how a suburb near you has grown.

A. **Write about the area where you live.**

1. Name the area where you live. _____

2. Is it a city, rural town, or suburb? _____

3. What is a nearby major city? _____

4. What are some nearby suburbs? _____

5. How do people travel between the suburbs and the city or between

   two suburbs? _____

6. Where do your family members work? _____

7. If they work outside the home, how do they travel to work? _____

8. Where do you go to school? _____

9. How do you get to school? _____

10. Where do you shop? _____

11. Where are the stores, and how do you get to them? _____

B. Think about the houses, roads, open land, stores, schools, and transportation in your suburb or a nearby suburban community. Are they new? Have they always been there? How have they changed? Write some questions you can ask someone to help you find out.

Features	Questions to Ask
Houses	
Roads	
Open Land	
Stores	
Schools	
Transportation	

C. Now interview an older family member, a neighbor, or a school employee. Ask your questions. Take notes on what you learn.

_____

_____

_____

_____

_____

_____

D. Decide how you will share your information. Then present what you learned about suburbs in your area.

_____ report _____ picture time line

_____ photographs, drawings _____ tape recording

Lesson 13

From Seed to Sandwich

As you read about the steps involved in making bread, think about how it shows that geography connects people and places.

Bread is a basic food in many parts of the world. It comes in an amazing variety of shapes and sizes. People can make it at home or buy it in a store. Let's look at the geography involved in the process of making bread—from seed to sandwich.

Plant seeds

Wheat grows on farms. Farmers plant spring wheat to harvest in fall and winter wheat to harvest the next spring or summer.

Harvest grain

When wheat is golden brown, it is ready to harvest. Farmers in the Midwest gather their crops by using a combine harvester. This machine is able to cut, thresh, and clean the grain.

Store grain

Trucks transport the wheat to the granary, a place where wheat is stored. Later, trucks take some of the wheat to grain elevators. Some grain is sold to milling companies. They grind the grain into flour.

TRANSPORT

Enjoy!

The seed that was planted in a farmer's field far away from here has become ready for you to enjoy at your own table.

TRANSPORT

Food for Thought

The longest loaf of bread on record was 3,491 feet and 9 inches long. It was baked at the Hyatt Regency Hotel in Guadalajara, Mexico, on January 6, 1991.

Buy bread

At the store the bread is put on shelves. When it comes to buying bread, people have many choices.

TRANSPORT

Package bread

After the bread cools, it is sliced and wrapped. The wrapped bread travels on big trucks to stores nearby and far away.

TRANSPORT

Make flour

At modern mills the milling process can take up to forty steps. Both white flour and whole wheat flour are made at the mill.

TRANSPORT

Bake bread

The flour is transported to a bakery. There, bakers add water, sugar, milk powder, yeast, vitamins, and minerals to the flour to make bread dough. A commercial bakery can produce many loaves of bread at one time.

Lesson 13

MAP SKILLS Using a Road Map to Get from Place to Place

We can use a road map to plan the best route to get where we want to go. The key to a road map can help us tell on what type of road we will be traveling. The map below shows one route that bread travels from seed to sandwich.

A. Look at the map of the Midwest. Circle the following.

1. Langdon, North Dakota
2. Duluth, Minnesota
3. Minneapolis, Minnesota
4. Sioux City, Iowa
5. Elmhurst, Illinois

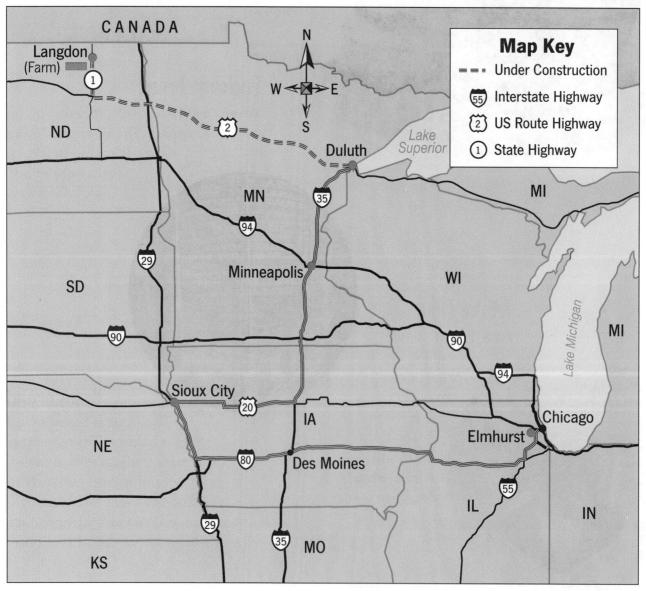

B. Study the map and answer the following.

1. Find the city of Langdon, North Dakota. Grain is harvested on a farm, then loaded onto a truck and taken to Langdon.

   a. What direction will the truck travel to get from Langdon to

   Highway 2? _____

   b. What road will the truck driver take to get from Langdon to

   Highway 2? _____

2. The grain is stored in Duluth, Minnesota.

   a. What direction will the truck travel to get to Duluth? _____

   b. Trace Highway 2 to Duluth. Why should the truck driver allow

   extra time? _____

3. The grain is sold to a mill in Minneapolis, Minnesota. There it will be ground into flour.

   a. What direction will the truck travel to get from Duluth to Minneapolis,

   Minnesota? _____

   b. What road will the truck driver use? _____

   c. What does the special symbol around the highway number mean? _____

   _____

   d. How would you give directions from Duluth to Minneapolis? _____

   _____

4. The flour is sold to a bakery in Sioux City, Iowa. There it will be made into bread. Write directions for getting from Minneapolis to Sioux City,

   Iowa. _____

5. Write directions for how the packaged bread will travel from the bakery in Sioux City, Iowa, to the grocery store in Elmhurst, Illinois.

   _____

Lesson 13

ACTIVITY
Find out the many places that products you use pass through.

A Link to Other Cities

The **BIG** Geographic Question

How do products you use connect you to other cities?

From the article you learned the steps it takes to make a loaf of bread. The map skills lesson showed you how people in different places can be connected by the foods they eat. Now find out how the products you use at home connect you to people in other places.

A. Write the names of four products you and your family use.

1. _____

2. _____

3. _____

4. _____

B. Read the information on the package for each product and use it to complete the table below.

Product Name	City or State of Distributor	Other Cities on Label

C. **Look at the city names you listed in the table.**

1. Copy each city and state name from your chart onto a small piece of scrap paper. Also include the product name on each piece of paper.
2. Locate a United States map in the Almanac. Find each city on the map. Tape your note on the map next to that city or state.
3. Use yarn or string to connect the cities to your hometown. Use tape to hold the strings in place.
4. Look at the map with your labels in place and answer the following questions.

    a. What product is produced closest to your town?

    _____

    b. What product is produced farthest from your town?

    _____

    c. What city, if any, provides more than one product?

    _____

D. **Select one of the products you listed and answer the following questions.**

1. What kind of product is it? Is it a food product, a clothing item, a

    school supply, etc.? _____

    _____

2. Do you think it is grown on a farm or made in a factory? _____

    _____

3. Do you think one person grows or makes it in one place or many

    people grow or help make it in many places? _____

    _____

E. **Do some further research on your selected product. Check your answers to the questions above and get more information about the product. Use the article as a model to create a step-by-step diagram of how your product is made.**

Lesson 14

A RACE FOR LIFE

As you read about how the Iditarod Trail Sled Dog Race started, think about how geography can affect a community.

The Iditarod Trail Sled Dog Race is more than 1,000 miles long. In Alaska's frozen north, racers run through forests, cross mountains and ice, and travel along the coast. The first race, in 1973, took twenty days. Today the racers finish in about ten days. The race is about more than speed. It is about keeping communities connected. It is about remembering others who raced the Iditarod Trail to save lives.

In January 1925, two children in Nome, Alaska, had a deadly disease called diphtheria. The doctor knew many people could die from the disease. He needed a special serum to save everyone. But there was a big problem. He needed to find the serum. Nome was an **isolated** town, meaning it was cut off from other towns in Alaska. It was also winter in Nome. No supplies could come in until the weather got better. Desperate, the doctor asked for help. The mayor sent telegrams to many places, trying to locate the serum. Finally, the serum was found in Anchorage.

Iditarod racers and their dogs brave snow on their trail from Anchorage to Nome.

There was another big problem. How would the serum get from Anchorage to Nome? It could not come by boat. The Bering Sea was frozen. It could not come by plane. At that time, airplanes had open cockpits, and a pilot might freeze to death. The only hope was for the serum to travel by train and dogsled. The journey would be very dangerous.

Dogsled relay teams met along the Iditarod Trail. The drivers and their dogs faced challenges. Racing day and night, they battled blizzards. They fought temperatures far below zero. They faced hungry, wild animals. Once, the serum was almost lost when it fell off the sled.

Despite these problems, the teams completed the journey in five days! The people of Nome were saved. Today the 1,159-mile Iditarod race is the toughest race in the world. Every year people come to test their skills and to remember the brave people and dogs who completed the first race for life.

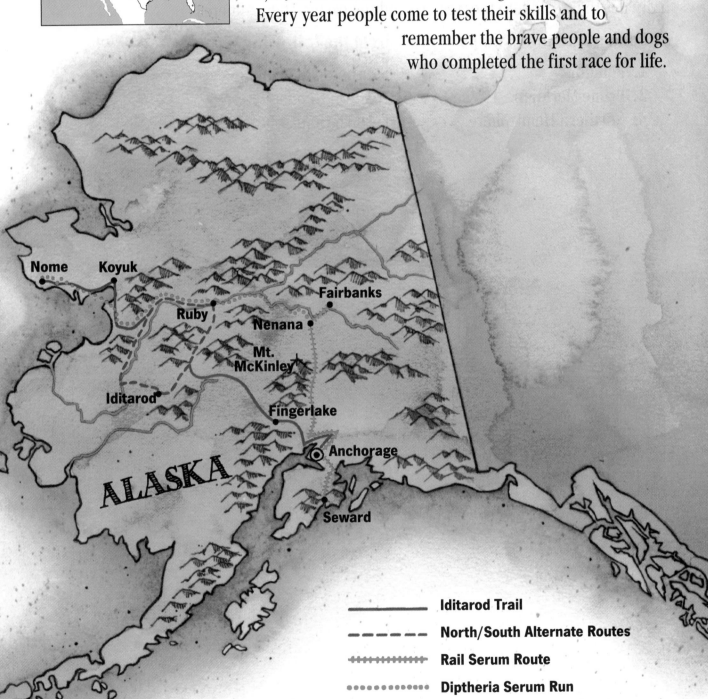

	Iditarod Trail
	North/South Alternate Routes
	Rail Serum Route
	Diptheria Serum Run

Lesson 14

MAP SKILLS
Using the Equator and Prime Meridian to Identify Hemispheres

Maps show imaginary lines that can help us explain where places are located. One line is the **equator**. It goes around Earth's center. It divides the world into the northern and the southern hemispheres. Another line is the **prime meridian.** It goes around Earth from top to bottom. It divides the world into the western and the eastern hemispheres.

A. Look at the world maps below. Circle the following terms.

1. Equator
2. Prime Meridian
3. Northern Hemisphere
4. Southern Hemisphere
5. Western Hemisphere
6. Eastern Hemisphere

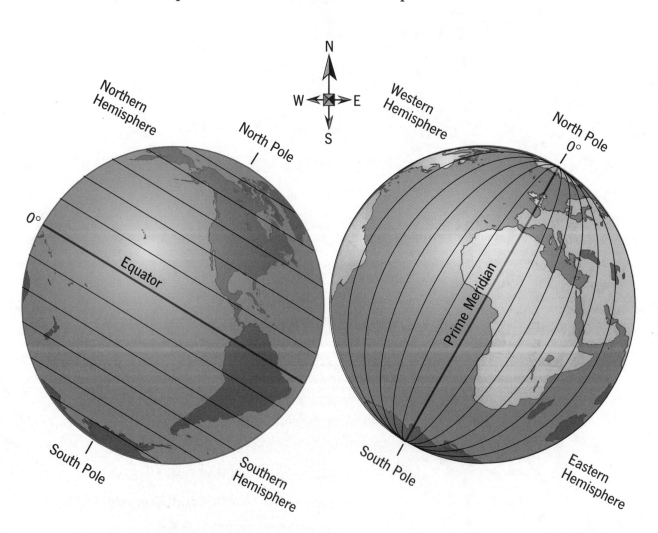

B. Look at the map with the equator labeled.

1. In what direction would you move your finger to trace the equator?

   _____

2. What is the half of Earth north of the equator called?

   _____

3. What is the half of Earth south of the equator called?

   _____

4. Is the Iditarod Trail in the northern hemisphere or the southern hemisphere?

   _____

C. Look at the map with the prime meridian labeled.

1. In what direction would you move your finger to trace the prime

   meridian? _____

2. What is the half of Earth to the west of the prime meridian called?

   _____

3. What is the half of Earth to the east of the prime meridian called?

   _____

4. Is the Iditarod Trail in the western hemisphere or the eastern hemisphere?

   _____

D. Find the continent and country of your home state on one of the maps.

1. Is your home state in the northern or southern hemisphere? _____

   _____

2. Is it in the western or eastern hemisphere? _____

3. Which is closer to the equator, your home state or Alaska? _____

   _____

Lesson 14

ACTIVITY Find out where other isolated
communities are located.

Community Connections

The **BIG**
Geographic Question

**Why do isolated communities
need to stay connected with
other communities?**

**From the article you learned about the Iditarod Trail Sled Dog
Race, and how the Iditarod Trail connected an isolated community
with other places. The map skills lesson showed you how to locate
communities on a part of the globe. Now you will find out about
other isolated communities and how they connect with other
people and places.**

A. Circle one of these isolated communities to read about.

   1. Buffalo, New York, in the 1800s

   2. Salt Lake City, Utah, in the 1800s

B. Answer as many of the following questions as you can about
the isolated community you selected. Look in the Almanac
for help with answering these questions.

   1. Why did people settle there? _____

   2. Why was it isolated? _____

   3. Why might the community need to be in touch with other communities?

   _____

   4. Did any specific event make that community build a connection with

      other communities? _____

      _____

C. Answer the following questions, then discuss them with a classmate.

1. How do you think people decide where to build communities? _____

_____

_____

2. What physical features might isolate a community? _____

_____

_____

3. Why do you think people in communities need to be in touch with

   others? _____

_____

_____

4. How might people in an isolated community connect with other

   communities? _____

_____

_____

D. Now create a collage that represents the isolated community you selected to learn about. You may draw, cut out, or paint pictures and paste them on your collage. It should help viewers understand how the community was isolated and how it later became connected to other communities. Share your collage with the class and compare it with your classmates' collages.

WHAT'S THE CONNECTION?

As you read about the advances in transportation and communication, think about the role geography played in these advances through time.

There was a time when it took days, weeks, and even months for people to travel between communities. Today we can fly across the country in less than a day. We can send and receive news within seconds. What changes have made this possible? What geographical obstacles had to be overcome?

Hundreds of years ago, Native Americans traveled by foot or water. When they needed to send messages long distances, they sent smoke signals or used drums. The early colonists traveled on foot, by water, or on horseback. It wasn't until much later that roads were cut through thick forests, and stagecoaches carried passengers between cities and towns.

foot travel

CHANGES IN TRANSPORTATION

covered wagon

train

paddle wheel steamers

zeppelin

With the invention of trains in the mid-1800s, overland travel became even easier. Mountains were no longer an obstacle because railroad tracks could be made to go over, around, and through them. By the 1920s cars became the chief means of transportation in the United States. In the 1950s airplanes made travel even faster.

CHANGES IN COMMUNICATION

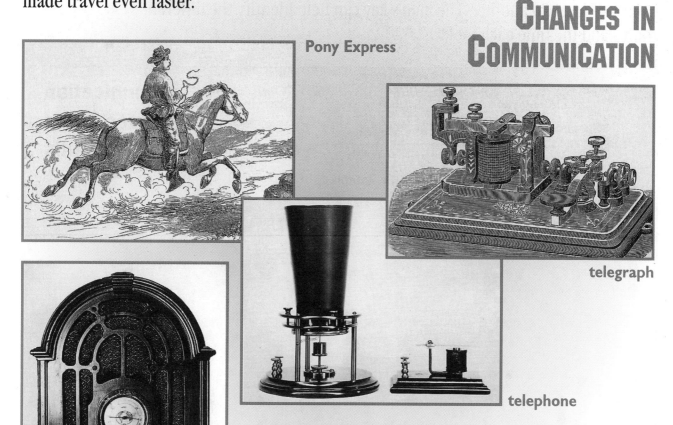

Pony Express

telegraph

telephone

radio

A BLAST FROM THE PAST
You already know what a car, subway, plane, and computer look like. Check out these images of transportation methods and communication tools from the past.

New ways of communicating have changed the way people live and do things, too. Centuries ago, news traveled only as fast as people did—on foot, horseback, or ship. Messages were sent across land by stagecoach and the Pony Express over a period of days or weeks. When a system of telegraph wires was connected across the country, messages could be sent in minutes.

The invention of the telephone improved communication even further in the late 1800s. Radio and television, which did not need wires to send messages, came along later and changed communication even more. Today millions of computers are connected by the Internet, an information "superhighway" used for sending and receiving information instantly using telephone lines.

MAP SKILLS Using a Map to Figure Travel Time

Maps can be used to show the different methods used to send messages from one place to another. The map's key can help identify the different routes and the time each took.

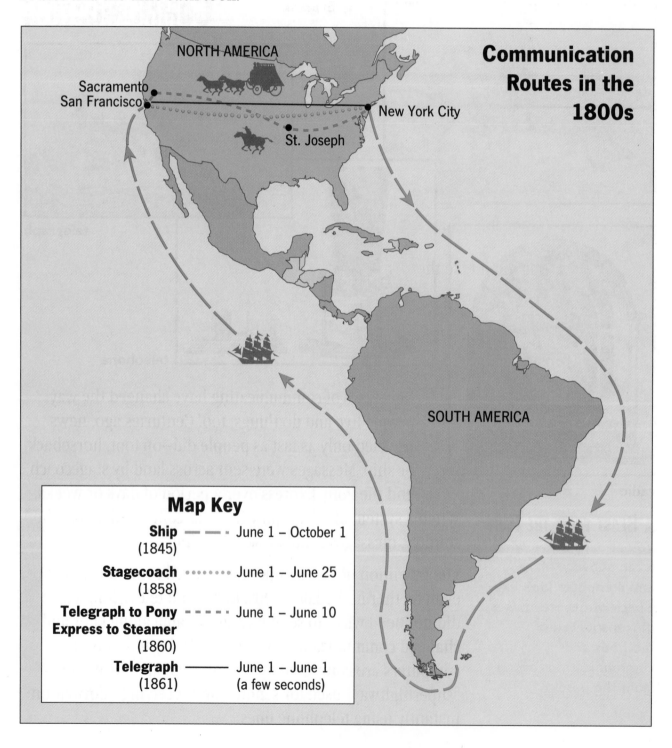

Communication Routes in the 1800s

NORTH AMERICA

Sacramento
San Francisco

New York City

St. Joseph

SOUTH AMERICA

Map Key

Ship — — — June 1 – October 1
(1845)

Stagecoach •••••••• June 1 – June 25
(1858)

Telegraph to Pony – – – – June 1 – June 10
Express to Steamer
(1860)

Telegraph ———— June 1 – June 1
(1861)   (a few seconds)

A. Look at the map and answer the following questions.

1. What color and symbol show the route of the stagecoach? _____

2. What color and symbol show the route of the ship? _____

B. Answer the following questions about each route on the map.

1. Look at the stagecoach route.
   a. What is the starting date? _____

   b. When did it get to San Francisco? _____

   c. How long did it take the stagecoach to deliver the message from

   New York to San Francisco? _____

2. Look at the ship route.
   a. What is the starting date? _____

   b. When did the ship get to San Francisco? _____

   c. How long did it take the ship to deliver the message? _____

3. Looking at the message route in 1860, answer the following
   questions.
   a. What three methods were used to send a message?

   _____   _____   _____

   b. How long did it take to deliver the message during the 1860 trip

   using these three methods? _____

   c. What was the time difference between how long it took in 1858

   and in 1860? _____

4. Find the delivery method that was fastest.
   a. How was it sent and in what year was the method used?

   _____

   b. How long did it take? _____

Lesson 15

ACTIVITY

Show the time it takes to send messages in different ways.

Message Express!

The **BIG** Geographic Question

What effect did geography have on the amount of time it took to send messages in the mid-1800s?

The article helped you understand how methods of communication and transportation changed over time. The map skills lesson showed you that developments in communication shortened the time needed to send messages. Now make a graph to show what you have learned.

A. List some ways of sending messages that you learned about in the article on pages 86–87.

1. _____   5. _____

2. _____   6. _____

3. _____   7. _____

4. _____   8. _____

B. List the ways that were shown in the map skills lesson.

1. _____

2. _____

3. _____

4. _____

5. _____

C. Use what you've learned to fill in the chart.

Year	Type of Communication	Approximate Time to Deliver Message
1845		
1858		
1860		
1861		

D. Organize your information on a graph. Show the difference in time required to send a message from New York City to San Francisco at different points in time.

Number of Days

```
100 ─┼─
 90 ─┼─
 80 ─┼─
 70 ─┼─
 60 ─┼─
 50 ─┼─
 40 ─┼─
 30 ─┼─
 20 ─┼─
 10 ─┼─
```

Date 1845   1858   1860   1861

Method _____ _____ _____ _____

Lesson 16

From "Ghost Town" to "Ski Town"

As you read about Telluride, think about how geography has changed the community over time.

Telluride, Colorado, is located high in the Rocky Mountains on the banks of the San Miguel River. It lies at the end of a **box canyon,** a valley so steep its walls seem almost vertical. Geography has played a key role in Telluride's history.

For years the town was called Columbia. Then, in 1875, silver and gold were discovered. People rushed there to mine the ore. The town was renamed Telluride in 1881 after a miner there found a metal called tellurium, which he thought was valuable.

Telluride turned out to be a town rich in natural resources, but it was isolated. Moving ore to mills in other towns was difficult and costly. By the end of the 1880s the mining stopped.

People left in the late 1880s, and Telluride became a ghost town.

The railroad connected Telluride to the outside world.

In 1890 the Rio Grande Southern Railroad reached Telluride. The ore could be transported to mills. The miners returned, and the town prospered. Its population rose to 4,000.

In 1914 a flood destroyed homes and businesses. By 1930, at the time of the Great Depression, the bank and most mines closed. The population fell again, this time to about 500.

In 1953 the Idarado Mining Company bought the mines. Telluride was revived as it mined millions of dollars worth of copper, lead, zinc, silver, and gold. In 1978 the ores ran out, and the Idarado Mining Company closed.

The mines no longer support the people of Telluride. Still, the community is thriving. The winter snow in the high mountains attracts skiers. In fall the mountains and the surrounding national forest draw many hikers to the area.

People come to Telluride by car or plane. Some are rowed down the San Miguel River in a gondola. The lofty peaks, once an obstacle, are now the main attraction for Telluride's 1,300 residents and its many visitors.

Skiers enjoy the snowy slopes of the Telluride ski area.

Lesson 16

MAP SKILLS Using a Map Scale to Figure Out Distance

Maps can show how far places are from each other. We can use a map scale to figure this out. On this map scale, one inch on a ruler would equal fifty miles on the map.

A. Look at the map of Colorado below. Circle the following items.

1. Telluride
2. Denver (capital of Colorado)
3. Roads leading to Telluride
4. Rivers near Telluride
5. Elevation of mountains near Telluride
6. The Rocky Mountains

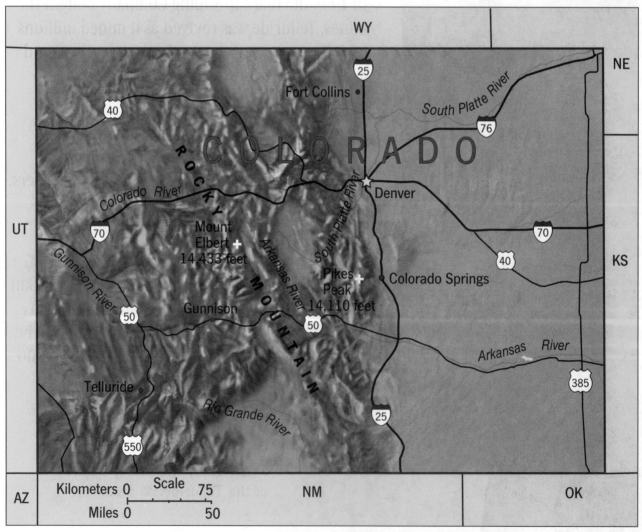

94

B. Find the distance between Telluride and Denver in air miles, or "as the crow flies."

1. With a ruler, measure the number of inches between Telluride and Denver on the map. Write the number of inches below.

    _____ inches

2. Use the map scale to find out the following.

    One inch stands for _____ miles.

3. Write down and multiply the numbers in questions 1 and 2 to find out the distance in miles.

    _____ X _____ = _____

4. On the map, Telluride is approximately _____ miles from Denver by air.

C. Road and river routes are often curvy, so they take a little more time to figure out.

1. Take a piece of string and bend it along a road route from Denver to Telluride.

2. Mark the string at Denver, then straighten it out.

3. Measure the string up to the mark. Write the number of inches below.

    _____ inches

4. Write down and multiply the number of inches of string by the number of miles an inch stands for.

    _____ X _____ = _____

5. Telluride is approximately _____ miles from Denver by car.

6. Write the directions for the road route you measured from Telluride to Denver. Be sure to include a description of any mountains or rivers you would cross.

    _____

    _____

Lesson 16

ACTIVITY
Find out why your community was settled where it is.

Your Community: Then and Now

The **BIG** Geographic Question	**Where and why was your community first settled, and how has it changed over time?**

From the article you learned why the community of Telluride was settled where it is and how it changed over time. The map skills lesson showed you how to use a map scale. Now find out how your community was settled and how it has changed.

A. What are two key physical features of your community?

1. _____

2. _____

B. Answer as many of the following questions about your community as you can.

1. When was your community settled? _____

_____

2. Who settled it? _____

_____

3. Why did they settle there? _____

_____

_____

4. Where does your community get its water? _____

C. What do you think your community was like when it was first settled? What is it like now? Write an *M* for *more,* an *L* for *less,* or *S* for *same* for the items listed in the chart.

	Before	Now
open land		
trees		
houses		
roads		
rivers, lakes, or creeks		
farms		
factories		

D. Talk to your friends and family. Go to the library. Find out whether your answers are correct! Then think about what your community is like now. How has it changed? Use the information to draw a picture or make a map of what your community was like when people first settled there and what it is like now.

Before	Now

Lesson 17

HOUSTON'S HORIZON

As you read about Houston, think about how the actions of humans affected its geography and the type of city it is.

When Houston, Texas, was first settled in 1836, it was a dusty little town 50 miles from the port city of Galveston. By 1990 it was the fourth largest U.S. city and our nation's third busiest port.

Houston was settled along Buffalo Bayou, which led to the Gulf of Mexico. The **bayou,** a creek or minor river, wasn't deep enough or wide enough for ships to use. This made it difficult for the people of Houston to travel to or to get goods from other places.

In the 1830s many farmers in Texas grew cotton. They needed a way to get their cotton to the port at Galveston to be shipped out. It wasn't until the 1850s, when railroad lines were built, that farmers were able to send their cotton from smaller towns to Houston by rail. It was then sent by barge along the Buffalo Bayou to Galveston. The railroad made it much easier for goods and people to get in and out of Houston.

Today skyscrapers line the streets of downtown Houston.

98

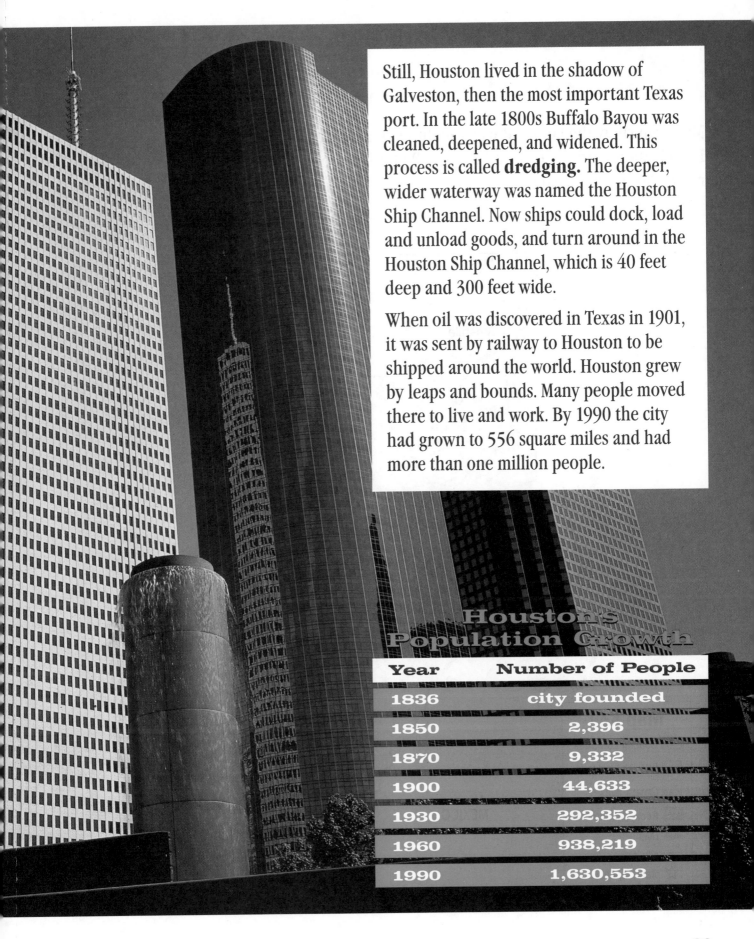

Still, Houston lived in the shadow of Galveston, then the most important Texas port. In the late 1800s Buffalo Bayou was cleaned, deepened, and widened. This process is called **dredging.** The deeper, wider waterway was named the Houston Ship Channel. Now ships could dock, load and unload goods, and turn around in the Houston Ship Channel, which is 40 feet deep and 300 feet wide.

When oil was discovered in Texas in 1901, it was sent by railway to Houston to be shipped around the world. Houston grew by leaps and bounds. Many people moved there to live and work. By 1990 the city had grown to 556 square miles and had more than one million people.

Houston's Population Growth

Year	Number of People
1836	city founded
1850	2,396
1870	9,332
1900	44,633
1930	292,352
1960	938,219
1990	1,630,553

MAP SKILLS
Using a Map to Identify the
Water Features of an Area

Maps can show the land and water features of an area. Land features
include mountains, plains, and deserts. Water features include
rivers, lakes, and oceans.

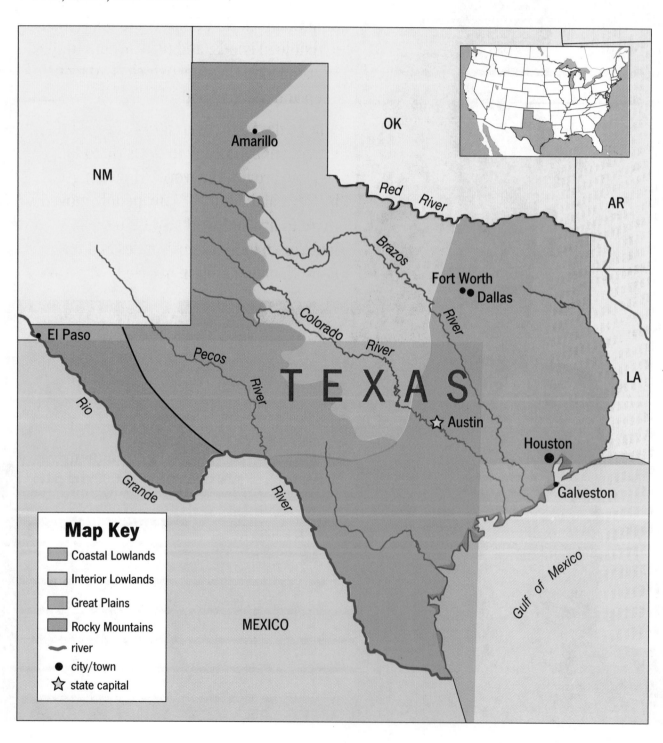

A. Look at the map of Texas. Circle the following.

    1. Houston

    2. Dallas

    3. Rio Grande

    4. Galveston

    5. Pecos River

    6. Gulf of Mexico

B. Use the map to complete the following.

    1. Look at the map key. Draw or write the name of the symbol or color that stands for each thing listed.

        a. The symbol for a river is _____.

        b. The symbol for a city is _____.

        c. The color that stands for the Great Plains region is _____.

    2. What is the capital of Texas? _____

    3. Find Austin on the map. On what river is Austin located?

        _____

    4. What river runs between Texas and Mexico?

        _____

    5. What city is located on the Rio Grande?

        _____

    6. Which city is located closest to the Gulf of Mexico?

        _____

    7. Why is it good for a city to be located on or near a river?

        _____

        _____

        _____

Lesson 17

ACTIVITY Find out the source of land and water features in your community.

Your Community's Land and Water

The **BIG**
Geographic Question

Which land and water features in your community are natural?

From the article you learned about the land and water features that helped Houston grow from a small town to the fourth largest city in the United States. In the map skills lesson you used a map to identify important waterways located in Texas. Now find out about land and water features in and around your community.

A. Most land and water features are **natural,** or made by nature. However, some are **artificial,** or built by people. Circle each artificial feature listed below.

river	dam	airport	forest
highway	mountain	canal	bridge
ocean	school	railroad	island

B. Complete the following items with information about your community.

1. List the most important land features of your community.

_____   _____

_____   _____

_____   _____

2. Put an *N* beside the above features that are *natural* and an *A* beside the ones that are *artificial.*

3. List the most important water features of your community.

_____   _____

_____   _____

_____   _____

4. Put an *N* beside the above features that are *natural* and an *A* beside the ones that are *artificial*.

C. Using the chart below, place a check mark in the box that shows each land and water feature found in your community and each found in Houston.

Land and Water Features	Your Community	Houston
lake		
river		
harbor		
mountain		
forest		
highway		
bridge		
airport		
railroad		

D. Compare your community with Houston. Think about how artificial features changed Houston. Then think about artificial features that have changed your community. Write a paragraph explaining whether you think the artifical features were a good or bad change for your community.

Islands of Fire and Rain

As you read about Hawaii, think about how its geography affects where and how people live on the islands.

Being an island really sets a place apart! Hawaii, with its active volcanoes, black beaches, and rain shadows, is very different from the mainland states.

The state of Hawaii is actually made up of eight islands. The largest of the eight islands is Hawaii. People call it the "Big Island." Like the other Hawaiian Islands, Hawaii is really the top of a huge underwater mountain created by the buildup of lava over three million years. The lava hardened and rose above the water level in the Pacific Ocean.

The "Big Island" still has two active volcanoes—Mauna Loa and Kilauea. Their lava has added more than 70 acres of land to Hawaii in the last ten years. Kilauea, the world's most active volcano, erupts every few months.

The climate on each side of Hawaii's high mountains can be very different. For example, Hilo, which is located on the east side of the "Big Island" is very wet. Kailua-Kona, located on the west side, is very dry. The reason for this is a process called **rain shadow.** Cool winds from the east side of the island produce rain that falls into the mountains. As the moisture clouds rise up the mountainside, they become drier. By the time they reach the west side of the mountains, they are dry.

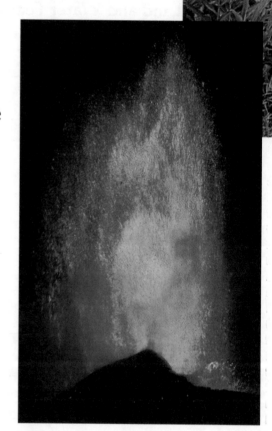

Spitting rocks and fire, Kilauea's fiery lava flows to the sea.

The east side of the mountains get more than 200 inches of rain each year. Farmers can grow sugar, macadamia nuts, and flowers in this area. Also, more people live on the east side of the island. However, most of the island's hotels and resorts are on the west side.

In the Hawaiian islands, the sun shines for up to 344 days a year, providing warm temperatures all year. In addition, its location near the equator and cool winds off the Pacific Ocean help keep the temperatures pleasant. Hawaii's beautiful weather and its many stunning sites make it a popular vacation spot.

The "Big Island" boasts beautiful mountains, valleys, cliffs, beaches, and waterfalls.

Wet ocean winds from the east cool and drop their moisture as rain in the mountains. The dry winds rush over the mountains, forming a "rain shadow" and leaving the west with little or no rain.

Dry air

mountain

mountain

Moist air

ocean

ocean

Lesson 18
MAP SKILLS Using Maps to Locate Places Using Directions

The large map below shows the eight major islands that make up the state of Hawaii. The small map is a locator map. It shows where Hawaii is in relation to the rest of the United States. The rest of the United States, which is one big piece of land, is called the **"mainland."**

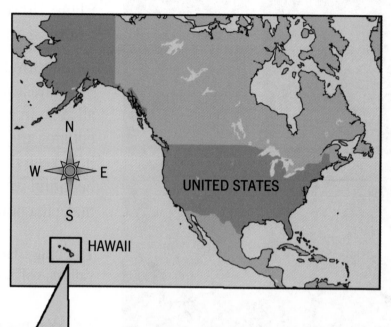

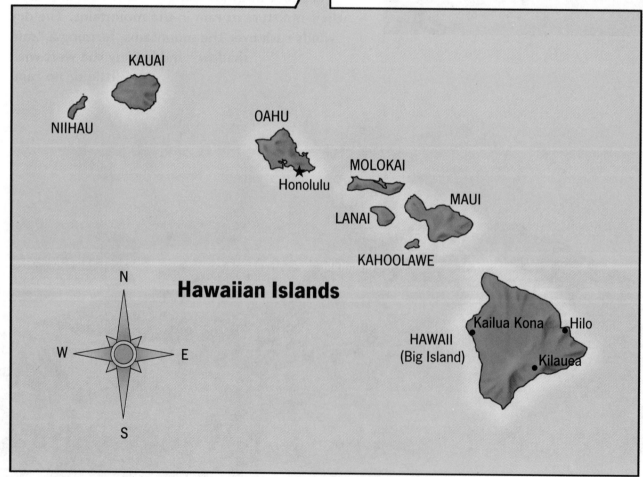

A. Look at the main map of Hawaii. Find and number the following places:

1. the "Big Island" Hawaii   **3.** Kailua-Kona   **5.** the island of Oahu

2. Hilo   **4.** Kilauea   **6.** the island of Maui

B. The compass rose usually shows only the cardinal directions—north, south, east, and west. Sometimes locations are between these four directions. Northeast, northwest, southeast, and southwest are intermediate directions. Fill in all eight directions on the compass rose below.

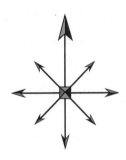

C. Use the compass rose to help with the following directions on the map.

1. Is Hawaii east or west of California?

_____

2. Choose a point on the west coast of the mainland. If you were going from this point to the state of Hawaii, in which direction would you be going?

_____

3. If you were going from the "Big Island" Hawaii to the west coast of the mainland, in which direction would you be going?

_____

4. If you were going from the "Big Island" Hawaii, to Honolulu, the capital of the state of Hawaii, in which direction would you be going?

_____

107

Lesson 18

 Plan a field trip to the island of Hawaii to study some of its natural wonders.

Exploring Hawaii

The **BIG** Geographic Question

What are some of Hawaii's unique natural features?

A. Suppose you were taking a field trip to the Big Island in January. You would need to plan transportation from where you live to the Big Island.

1. What form of transportation would you most likely take from where you live? _____

2. In which direction would you have to travel from your home to Hawaii? _____

B. Using information from the article and Almanac, list on the chart below some places in Hawaii that you would like to visit. Describe each place. One suggestion is shown on the chart for you.

Places to Visit	Description
Kilauea	the world's most active volcano

C. **Now think about the kinds of climate you will find in the places you plan to visit. Use the article and Almanac to find the following information.**

1. What is the average high temperature on Hawaii in January?

_____

2. What is the average low temperature on Hawaii in January?

_____

3. How does the climate on the east coast of Hawaii differ from the climate on the west coast of Hawaii?

_____

D. **Think about the clothing you will need to take with you. List and briefly explain why you will need each item.**

_____

_____

_____

_____

_____

_____

_____

E. **Imagine that you are in Hawaii. Using an index card, make a postcard to send home telling what you like about one of Hawaii's natural features. Draw a picture of the natural feature on the front of your card. Write your message on the back.**

ALMANAC

SEVEN CONTINENTS

Africa
Antarctica
Asia
Australia
Europe
North America
South America

THE FOUR OCEANS

Arctic
Atlantic
Indian
Pacific

THE FOUR BASIC LANDFORMS

Hills
Mountains
Plains
Plateaus

STATE NAMES	ABBREVIATIONS
Alabama	AL
Alaska	AK
Arizona	AZ
Arkansas	AR
California	CA
Colorado	CO
Connecticut	CT
Delaware	DE
Florida	FL
Georgia	GA
Hawaii	HI
Idaho	ID
Illinois	IL
Indiana	IN
Iowa	IA
Kansas	KS
Kentucky	KY
Louisiana	LA
Maine	ME
Maryland	MD
Massachusetts	MA
Michigan	MI
Minnesota	MN
Mississippi	MS
Missouri	MO

Montana	MT
Nebraska	NE
Nevada	NV
New Hampshire	NH
New Jersey	NJ
New Mexico	NM
New York	NY
North Carolina	NC
North Dakota	ND
Ohio	OH
Oklahoma	OK
Oregon	OR
Pennsylvania	PA
Rhode Island	RI
South Carolina	SC
South Dakota	SD
Tennessee	TN
Texas	TX
Utah	UT
Vermont	VT
Virginia	VA
Washington	WA
West Virginia	WV
Wisconsin	WI
Wyoming	WY

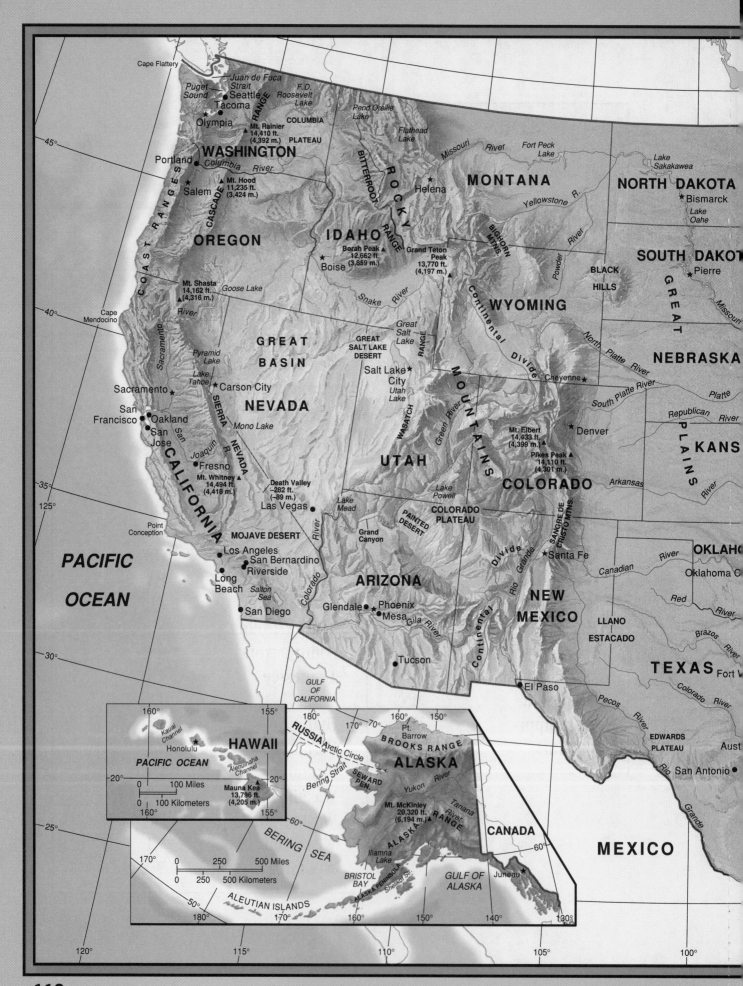

Cape Flattery
Juan de Fuca Strait
Puget Sound
Seattle
Tacoma
Olympia
Mt. Rainier 14,410 ft. (4,392 m.)
COLUMBIA
PLATEAU
F.D. Roosevelt Lake
Pend Oreille Lake
WASHINGTON
Portland
Columbia River
Flathead Lake
Missouri River
Fort Peck Lake
MONTANA
NORTH DAKOTA
Lake Sakakawea
★ Bismarck
Mt. Hood 11,235 ft. (3,424 m.)
Salem
OREGON
Helena ★
ROCKY
BITTERROOT
RANGE
CASCADE RANGE
Yellowstone R.
Lake Oahe
IDAHO
Borah Peak 12,662 ft. (3,859 m.)
Grand Teton Peak 13,770 ft. (4,197 m.)
BIGHORN MTNS.
Powder River
SOUTH DAKOTA
★ Pierre
COAST RANGES
Mt. Shasta 14,162 ft. (4,316 m.)
Goose Lake
Boise ★
Snake River
WYOMING
BLACK HILLS
GREAT
Cape Mendocino
Sacramento River
Pyramid Lake
GREAT BASIN
GREAT SALT LAKE DESERT
Great Salt Lake
Continental Divide
North Platte River
NEBRASKA
PLAINS
Lake Tahoe
Carson City ★
Salt Lake City ★
Utah Lake
Cheyenne ★
South Platte River
Platte
Sacramento ★
San Francisco
Oakland
San Jose
SIERRA
NEVADA
Mono Lake
NEVADA
WASATCH RANGE
MOUNTAINS
Green River
Mt. Elbert 14,433 ft. (4,399 m.)
★ Denver
Republican River
KANSAS
CALIFORNIA
San Joaquin River
Fresno ●
Mt. Whitney 14,494 ft. (4,418 m.)
Death Valley −282 ft. (−89 m.)
UTAH
Lake Powell
Pikes Peak 14,110 ft. (4,301 m.)
Arkansas River
Point Conception
Lake Mead
COLORADO PLATEAU
PAINTED DESERT
COLORADO
SANGRE DE CRISTO MTNS.
Las Vegas ●
Colorado River
MOJAVE DESERT
Los Angeles ●
San Bernardino
Riverside
Long Beach ●
Salton Sea
San Diego ●
Grand Canyon
Divide
Rio Grande
★ Santa Fe
Canadian River
OKLAHOMA
Oklahoma City
PACIFIC
OCEAN
ARIZONA
NEW MEXICO
LLANO
ESTACADO
Red River
Glendale ●
Phoenix ★
Mesa ●
Gila River
Continental Divide
Brazos River
Tucson ●
TEXAS
Fort W
El Paso ●
Pecos River
Colorado River
EDWARDS PLATEAU
Aust
GULF OF CALIFORNIA
Rio Grande
San Antonio ●

160° 155°
Kauai Channel
Honolulu ●
HAWAII
PACIFIC OCEAN
Alenuihaha Channel
0 100 Miles
0 100 Kilometers
Mauna Kea 13,796 ft. (4,205 m.)
160° 155°

180°
RUSSIA
Arctic Circle
Bering Strait
170° 70° 160° 150°
Pt. Barrow
BROOKS RANGE
SEWARD PEN.
ALASKA
Yukon River
60°
Tanana River
Mt. McKinley 20,320 ft. (6,194 m.)
ALASKA RANGE
CANADA
60°
Iliamna Lake
Shelikof Str.
Juneau ★
130°
BERING SEA
0 250 500 Miles
0 250 500 Kilometers
BRISTOL BAY
ALASKA PENINSULA
GULF OF ALASKA
ALEUTIAN ISLANDS

MEXICO

45°
125°
40°
35°
30°
25°

120° 115° 110° 105° 100°

112

CANADA

MAINE

Lake of the Woods

Red Lake

Lake Superior

Mt. Washington 6,288 ft.(1,905 m.)
Moosehead Lake

★Augusta

MICHIGAN

MINNESOTA

WISCONSIN

ADIRONDACK MTNS.

St. Lawrence River

Lake Champlain

Montpelier

N.H.

★Concord

VT.

Lake Huron

Lake Michigan

neapolis
★St. Paul

Mississippi River

Milwaukee
Madison

Grand Rapids

Lansing
Detroit

Rochester

Lake Ontario

Syracuse

Albany

NEW YORK

Hartford

Boston

MASS.

Cape Cod

★Providence

R.I.

Niagara Falls

Buffalo

Lake Erie

Susquehanna River

New Haven

CONN.

IOWA

Chicago
Gary
Hammond

Toledo

Cleveland

Akron
Canton

Youngstown

PENNSYLVANIA

Harrisburg

Pittsburgh

Newark

New York

N.J.

Philadelphia
Trenton

ILLINOIS

CENTRAL

LOWLAND

OHIO

Columbus

Dayton

Camden

Dover

MD.

DEL.

DELAWARE BAY

aha
★Des Moines

Springfield

★Indianapolis

INDIANA

Cincinnati River

Ohio R.

Wabash R.

Frankfort

Charleston

WEST VIRGINIA

Baltimore

Annapolis

Washington

D.C.

ka
Kansas City

East St. Louis
St. Louis

Louisville

KENTUCKY

Richmond

Newport News

Norfolk

CHESAPEAKE BAY

ATLANTIC

OCEAN

Jefferson City

MISSOURI

Harry S. Truman Res.

OZARK PLATEAU

VIRGINIA

Roanoke River

Cape Hatteras

nsas City

Tulsa

OZARK PLATEAU

R.S. Kerr Res.

ARKANSAS

Cumberland River

Knoxville

★Nashville

TENNESSEE

Mt. Mitchell 6,684 ft. (2,037 m.)

PLATEAU

★Raleigh

Winston-Salem

NORTH CAROLINA

Lake Eufaula

oma

Little Rock★

Memphis

Mississippi River

Tennessee River

Cumberland R.

Columbia

SOUTH CAROLINA

allas

Birmingham

Atlanta

Chattahoochee R.

GEORGIA

ALABAMA

Alabama R.

Montgomery

COASTAL

PLAIN

LOUISIANA

★Jackson

MISSISSIPPI

Sam Rayburn Reservoir

Toledo Bend Res.

Lake Pontchartrain

Baton Rouge★

Jacksonville

★Tallahassee

FLORIDA

APPALACHIAN MOUNTAINS

Houston

New Orleans

Orlando

Cape Canaveral

Tampa

St. Petersburg

UNITED STATES

⊛ National capital

★ State capital

● Major city

── International boundary

── State boundary

0   150   300 Miles

0   150   300 Kilometers

Projection: Albers Equal Area

GULF OF MEXICO

↑
N

Cape Sable

Lake Okeechobee

Miami

Key West

Straits of Florida

THE BAHAMAS

CUBA

45°

65°

40°

35°

70°

30°

25°

95°   90°   85°   80°   75°

SIA

RUSSIA

CHUKCHI
SEA

Bering Strait

BERING
SEA

RENCE I.

SEWARD
PEN.

PENINSULA

KODIAK I.

ALASKA (U.S.)

Mt. McKinley
20,320 ft.
(6,194 m.)

ALASKA RANGE

Anchorage

Fairbanks

Yukon River

Point
Barrow

BEAUFORT
SEA

ARCTIC OCEAN

North
+
Pole

80°

70°

80°

70°

ARCTIC OCEAN

Nares Str.

ELLESMERE
ISLAND

QUEEN ELIZABETH
ISLANDS

VICTORIA
ISLAND

Great Bear
Lake

Mackenzie

MACKENZIE
MOUNTAINS

Mt. Logan
19,850 ft.
(6,050 m.)

Whitehorse

COAST MOUNTAINS

Juneau

GULF OF
ALASKA

ALEXANDER
ARCHIPELAGO

QUEEN
CHARLOTTE
ISLANDS

VANCOUVER
ISLAND

PACIFIC
OCEAN

Victoria

Vancouver

Seattle

Portland

Spokane

Snake

Columbia R.

Fraser
R.

CASCADE RANGE

COLUMBIA
PLATEAU

ROCKY

CANADA

Great Slave
Lake

Great Slave
River

Athabasca

Peace
R.

Lake
Athabasca

Edmonton

Calgary

North
Saskatchewan

South

Reindeer
Lake

Churchill
R.

Nelson
R.

Regina

Lake
Winnipeg

Winnipeg

Lake
Manitoba

Churchill

CANADIAN
SHIELD

GREAT

UNITED STATES

M

KALAALLIT NUNAAT
(GREENLAND)
(DENMARK)

ICE

Denmark Strait

Arctic Circle

Davis Strait

BAFFIN
BAY

BAFFIN
ISLAND

Hudson Strait

HUDSON
BAY

UNGAVA
PEN.

LABRADOR

LABRADOR
SEA

Smallwood
Res.

Cape F

NEWF

NEWFO

Lake
Superior

Lake
Huron

Lake Ontario

Quebec

Montreal

Ottawa

GULF OF
ST. LAWRENCE

Lawrence

R.

Halifax

Cape Sable

ST. LAWRENCE ST
(F

114

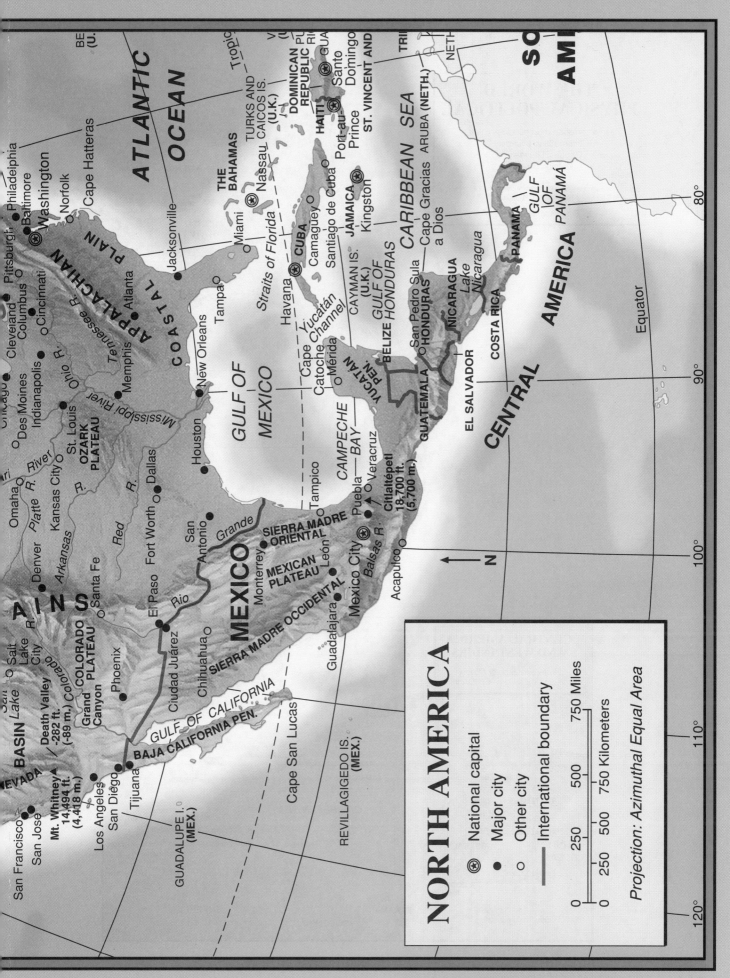

ATLANTIC OCEAN

BE (U.

Philadelphia
Baltimore
Washington
Norfolk
Cape Hatteras

APPALACHIAN

COASTAL PLAIN

Jacksonville

Cleveland
Columbus
Cincinnati
Pittsburgh

Atlanta
Tennessee R.
Memphis

Indianapolis
Des Moines
St. Louis
OZARK PLATEAU
Ohio R.

Omaha
Platte R.
Kansas City
Missouri River

Red R.

Arkansas

Denver
Santa Fe
COLORADO PLATEAU
Colorado R.

AINS
R

NEVADA
Salt Lake City
Salt Lake

Grand Canyon
Death Valley
-282 ft.
(-89 m.)

BASIN
San
Lake

Mt. Whitney
14,494 ft.
(4,418 m.)

San Francisco
San Jose
Los Angeles
San Diego
Tijuana

El Paso
Ciudad Juárez

Phoenix

Chihuahua

Fort Worth
Dallas

Rio
Grande

San Antonio

New Orleans

Tampico

Houston

GULF OF MEXICO

Tampa

Tropic

THE BAHAMAS

Nassau

Straits of Florida

Miami

Havana

CUBA

Camagüey

Santiago de Cuba

JAMAICA
Kingston

CAYMAN IS.
(U.K.)

Cape Catoche

Yucatán Channel

Mérida

YUCATÁN PEN.

CAMPECHE BAY

BELIZE

GUATEMALA

EL SALVADOR

HONDURAS
San Pedro Sula

NICARAGUA
Lake
Nicaragua

COSTA RICA

Cape Gracias
a Dios

CARIBBEAN SEA

ARUBA (NETH.)

PANAMA

GULF OF PANAMÁ

CENTRAL
AMERICA

PANAMÁ

Equator

TURKS AND
CAICOS IS.
(U.K.)

DOMINICAN
REPUBLIC

HAITI
Port-au-
Prince

Santo
Domingo

PU
RI

GUA

ST. VINCENT AND

TRI

NETH

SO
AM

Veracruz

Citlaltépetl
18,700 ft.
(5,700 m.)

Puebla

Mexico City

Balsas R.

Acapulco

León

MEXICAN
PLATEAU

SIERRA MADRE
ORIENTAL

MEXICO

Guadalajara

Monterrey

SIERRA MADRE OCCIDENTAL

GULF OF CALIFORNIA

BAJA CALIFORNIA PEN.

Cape San Lucas

GUADALUPE I.
(MEX.)

REVILLAGIGEDO IS.
(MEX.)

N

80°

90°

100°

110°

120°

NORTH AMERICA

⊛ National capital
● Major city
○ Other city
— International boundary

750 Miles
750 Kilometers
500
250
500
250
0
0

Projection: Azimuthal Equal Area

115

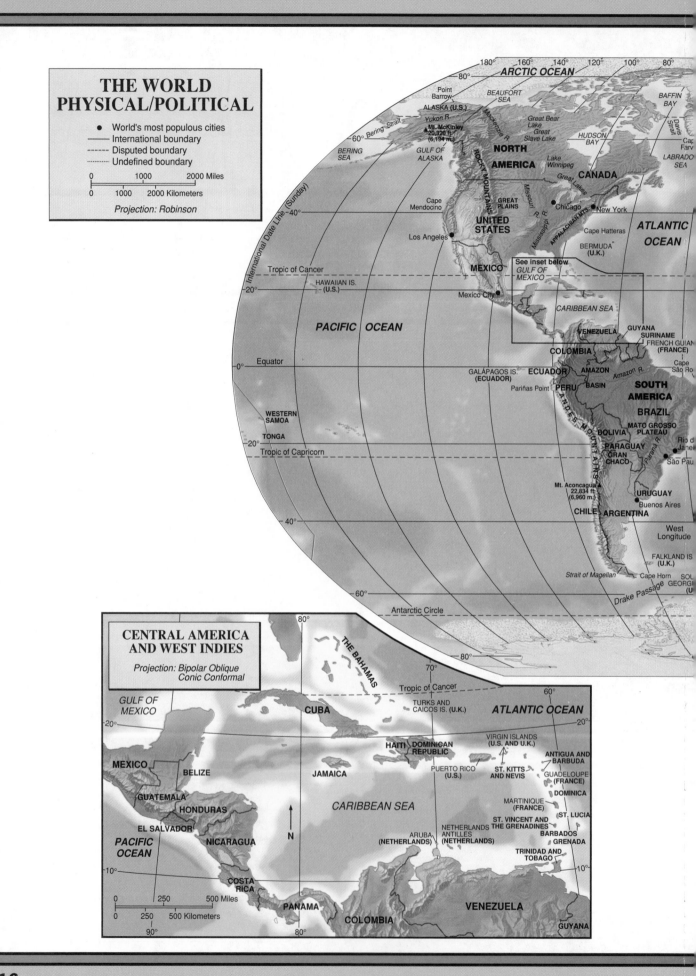

THE WORLD PHYSICAL/POLITICAL

- World's most populous cities
— International boundary
‒‒‒ Disputed boundary
⋯⋯ Undefined boundary

0   1000   2000 Miles
0   1000   2000 Kilometers

Projection: Robinson

ARCTIC OCEAN

180° 160° 140° 120° 100° 80°

80°

Point Barrow
BEAUFORT SEA

ALASKA (U.S.)
Yukon R.
Mt. McKinley
20,320 ft.
(6,194 m.)

BAFFIN BAY

Great Bear Lake
Great Slave Lake

HUDSON BAY

Cape Farv

DAVIS Strait

LABRADOR SEA

60° Bering Strait
BERING SEA

GULF OF ALASKA

ROCKY MOUNTAINS

NORTH AMERICA

CANADA

Lake Winnipeg

Great Lakes

Missouri R.

Mississippi R.

Chicago
New York

40°

Cape Mendocino

UNITED STATES

GREAT PLAINS

APPALACHIAN MTS.

Cape Hatteras

ATLANTIC OCEAN

Los Angeles

MEXICO

BERMUDA (U.K.)

See inset below
GULF OF MEXICO

Tropic of Cancer

20°

HAWAIIAN IS. (U.S.)

Mexico City

CARIBBEAN SEA

VENEZUELA

GUYANA

SURINAME

FRENCH GUIANA (FRANCE)

COLOMBIA

0° Equator

GALÁPAGOS IS. (ECUADOR)

ECUADOR

AMAZON

Amazon R.

Cape São Ro

PACIFIC OCEAN

PERU

BASIN

SOUTH AMERICA

Pariñas Point

BRAZIL

WESTERN SAMOA

BOLIVIA

MATO GROSSO PLATEAU

Rio d Janei

TONGA

20° Tropic of Capricorn

PARAGUAY
GRAN CHACO

Paraná R.

São Pau

URUGUAY

Mt. Aconcagua
22,834 ft.
(6,960 m.)

Buenos Aires

CHILE ARGENTINA

40°

West Longitude

FALKLAND IS. (U.K.)

International Date Line (Sunday)

Strait of Magellan
Cape Horn

SOU GEORGI (U

60°

Drake Passage

Antarctic Circle

80°

See inset below

CENTRAL AMERICA AND WEST INDIES

Projection: Bipolar Oblique Conic Conformal

80°

THE BAHAMAS

70°

60°

Tropic of Cancer

GULF OF MEXICO

CUBA

TURKS AND CAICOS IS. (U.K.)

ATLANTIC OCEAN

20°

20°

MEXICO

BELIZE

HAITI

DOMINICAN REPUBLIC

VIRGIN ISLANDS (U.S AND U.K.)

ANTIGUA AND BARBUDA

JAMAICA

PUERTO RICO (U.S.)

ST. KITTS AND NEVIS

GUADELOUPE (FRANCE)

GUATEMALA

CARIBBEAN SEA

DOMINICA

HONDURAS

MARTINIQUE (FRANCE)

ST. LUCIA

EL SALVADOR

ST. VINCENT AND THE GRENADINES

BARBADOS

PACIFIC OCEAN

NICARAGUA

N

ARUBA (NETHERLANDS)

NETHERLANDS ANTILLES (NETHERLANDS)

GRENADA

TRINIDAD AND TOBAGO

10°

10°

COSTA RICA

0   250   500 Miles
0   250   500 Kilometers

PANAMA

VENEZUELA

90°

80°

COLOMBIA

GUYANA

116

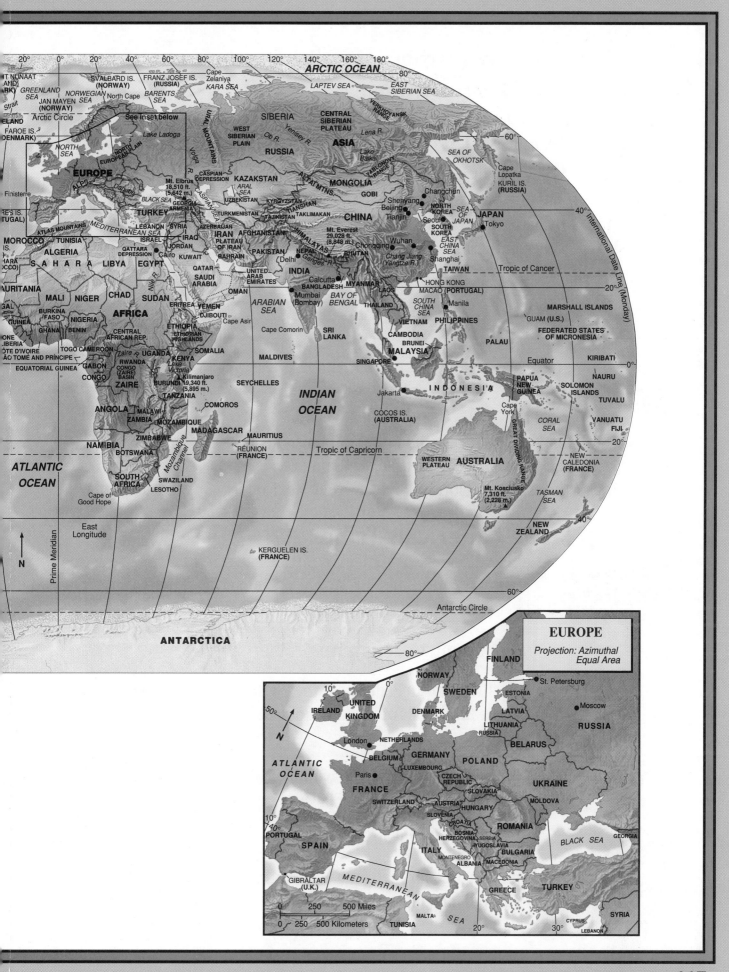

20° 0° 20° 40° 60° 80° 100° 120° 140° 160° 180°

ARCTIC OCEAN

T NUNAAT
AND
RK)

GREENLAND
SEA
JAN MAYEN
(NORWAY)
NORWEGIAN
SEA

Cape
Zelaniya
KARA SEA

LAPTEV SEA

EAST
SIBERIAN SEA

SVALBARD IS.
(NORWAY)

FRANZ JOSEF IS.
(RUSSIA)

BARENTS
SEA

North Cape

Strait

ELAND

FAROE IS.
DENMARK)

Arctic Circle

See Inset below

NORTH
SEA

SIBERIA

WEST
SIBERIAN
PLAIN

CENTRAL
SIBERIAN
PLATEAU

VERKHOYANSK

Lake Ladoga

Yenisey R.

Lena R.

RANGE

Finisterre

EUROPE

ALPS

Danube

Volga R.

URAL MOUNTAINS

Ob R.

RUSSIA

ASIA

Lake
Baikal

SEA OF
OKHOTSK

60°

Cape
Lopatka
KURIL IS.
(RUSSIA)

CASPIAN
DEPRESSION

KAZAKSTAN

MONGOLIA

Mt. Elbrus
18,510 ft.
(5,642 m.)

ARAL
SEA

ALTAI MTNS.

GOBI

Changchun

SEA
OF
JAPAN

RES IS.
TUGAL)

BLACK SEA

GEORGIA
ARMENIA

UZBEKISTAN

KYRGYZSTAN

TIANSHAN

Shenyang
Beijing

NORTH
KOREA

JAPAN

40°

TURKEY

CASPIAN SEA

TURKMENISTAN

TAJIKISTAN

TAKLIMAKAN

CHINA

Tianjin

Seoul
SOUTH
KOREA

Tokyo

MOROCCO

LEBANON
ISRAEL

SYRIA

IRAQ

AZERBAIJAN

AFGHANISTAN

Mt. Everest
29,028 ft.
(8,848 m.)

HIMALAYAS

Chongqing

Wuhan

EAST
CHINA
SEA

ATLAS MOUNTAINS

MEDITERRANEAN SEA

JORDAN

IRAN
PLATEAU
OF IRAN

PAKISTAN

NEPAL

BHUTAN

Chang Jiang
(Yangtze R.)

Shanghai

ALGERIA

TUNISIA

QATTARA
DEPRESSION

Cairo

KUWAIT

BAHRAIN

Delhi

Ganges R.

TAIWAN

Tropic of Cancer

20°

HARA

OCCO)

LIBYA

EGYPT

Nile R.

QATAR

SAUDI
ARABIA

UNITED
ARAB
EMIRATES

INDIA

Calcutta

BANGLADESH

MYANMAR

HONG KONG

MACAO (PORTUGAL)

URITANIA

MALI

NIGER

CHAD

SUDAN

ERITREA

YEMEN

OMAN

Mumbai
(Bombay)

ARABIAN
SEA

BAY OF
BENGAL

LAOS

THAILAND

SOUTH
CHINA
SEA

Manila

MARSHALL ISLANDS

GAL

GUINEA

BURKINA
FASO

NIGERIA

BENIN

AFRICA

CENTRAL
AFRICAN REP.

ETHIOPIA

DJIBOUTI

Cape Asir

Cape Comorin

SRI
LANKA

VIETNAM

CAMBODIA

PHILIPPINES

GUAM (U.S.)

FEDERATED STATES
OF MICRONESIA

ONE

LIBERIA

ÔTE D'IVOIRE

GHANA

TOGO

CAMEROON

ETHIOPIAN
HIGHLANDS

MALDIVES

BRUNEI

PALAU

KIRIBATI

ÂO TOME AND PRÍNCIPE

EQUATORIAL GUINEA

GABON

Zaire

UGANDA

KENYA

SOMALIA

MALAYSIA

SINGAPORE

Equator

0°

NAURU

CONGO

RWANDA
CONGO
(ZAIRE)
BASIN

Lake
Victoria

Mt. Kilimanjaro
19,340 ft.
(5,895 m.)

SEYCHELLES

INDONESIA

Jakarta

PAPUA
NEW
GUINEA

SOLOMON
ISLANDS

TUVALU

ZAIRE

BURUNDI

TANZANIA

INDIAN

OCEAN

Cape
York

CORAL
SEA

VANUATU

FIJI

ANGOLA

ZAMBIA

MALAWI

MOZAMBIQUE

COMOROS

MADAGASCAR

MAURITIUS

COCOS IS.
(AUSTRALIA)

GREAT DIVIDING RANGE

20°

ZIMBABWE

NAMIBIA

BOTSWANA

Mozambique Channel

RÉUNION
(FRANCE)

Tropic of Capricorn

WESTERN
PLATEAU

AUSTRALIA

NEW
CALEDONIA
(FRANCE)

**ATLANTIC
OCEAN**

SOUTH
AFRICA

SWAZILAND

LESOTHO

Cape of
Good Hope

Mt. Kosciusko
7,310 ft.
(2,228 m.)

TASMAN
SEA

40°

East
Longitude

Prime Meridian

N

KERGUELEN IS.
(FRANCE)

NEW
ZEALAND

60°

Antarctic Circle

80°

ANTARCTICA

International Date Line (Monday)

80°

40°

20°

60°

EUROPE

*Projection: Azimuthal
Equal Area*

10° 0°

FINLAND

NORWAY

SWEDEN

St. Petersburg

ESTONIA

Moscow

50°

UNITED
KINGDOM

IRELAND

DENMARK

LATVIA

RUSSIA

N

London

NETHERLANDS

LITHUANIA

RUSSIA

BELARUS

**ATLANTIC
OCEAN**

BELGIUM

GERMANY

POLAND

Paris

LUXEMBOURG

CZECH
REPUBLIC

UKRAINE

FRANCE

SWITZERLAND

SLOVAKIA

MOLDOVA

AUSTRIA

HUNGARY

SLOVENIA

CROATIA

ROMANIA

10°

40°

PORTUGAL

BOSNIA
HERZEGOVINA

SERBIA

YUGOSLAVIA

BLACK SEA

GEORGIA

SPAIN

ITALY

MONTENEGRO

BULGARIA

ALBANIA

MACEDONIA

GIBRALTAR
(U.K.)

MEDITERRANEAN

GREECE

TURKEY

0 250 500 Miles

MALTA

SEA

CYPRUS

SYRIA

0 250 500 Kilometers

TUNISIA

LEBANON

20° 30°

Time Line of Native American and European Settlements in North America

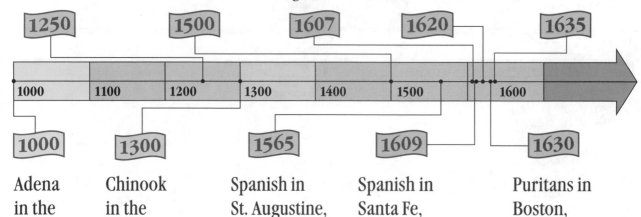

- Iroquois in the Northeast
- Sioux in the Great Plains
- Cherokee in the Southeast

Anasazi, or Pueblo, in the Southwest

English in Jamestown, Virginia

Pilgrims in Plymouth, Massachusetts

Dutch in New York City, New York

1250   **1500**   **1607**   **1620**   **1635**

1000   1100   1200   1300   1400   1500   1600

1000   **1300**   **1565**   **1609**   **1630**

Adena in the Midwest

Chinook in the Northwest

Spanish in St. Augustine, Florida

Spanish in Santa Fe, New Mexico

Puritans in Boston, Massachusetts

Source: *The World Book Encyclopedia,* 1991

Regional Native American Tribes

Great Plains Native Americans	Iowa, Kansa, Missouri, Wichita, Comanche, Omaha, Crow, Cheyenne, Sioux, Lakota
Spotlight on: Cheyenne Tribe	
Region	North American Plains near the Platte and Arkansas rivers; in the Black Hills of South Dakota; tall grasslands and rivers
Climate	warm, dry summers; cold winters with blowing snow
Shelter	tepees were made of long poles and buffalo skins which could be easily taken apart and moved
Resources	buffalo used for fuel, food, and clothing; fertile river valley allowed for some farming

Northeast Native Americans (Eastern Woodlands)	Mohawk, Cayuga, Seneca, Oneida, Onondaga, Iroquois, Delaware

Spotlight on: Seneca Tribe	
Region	western New York and eastern Ohio
Climate	warm summers; cold, snowy winters
Shelter	longhouses built of wood and bark
Resources	fertile soil and an abundance of fresh water from the Great Lakes allowed them to cultivate corn and other vegetables; hunted a variety of wild game

Northwest Native Americans	Makah, Haida, Okanagoh, Quinault, Nootka, Chinook, Spokane, Kalapuya, Kalispel, Shuswap

Spotlight on: Makah Tribe	
Region	Northwest Pacific coast of North America; Vancouver Island at northwest tip of Washington state
Climate	mild temperatures due to Pacific Ocean air currents which bring heavy rainfall
Shelter	wooden plank houses
Resources	forests for shelter; salmon, whale, and caribou used for food

Southeast Native Americans	Cherokee, Chickasaw, Choctaw, Creek, Seminole, Powhatan, Natchez, Timicua, Sauk, Caddo

Spotlight on: Cherokee Tribe	
Region	southern Appalachian region; North Carolina and Tennessee
Climate	mild seasons; warm summers and wet winters
Shelter	wigwams: huts with arched frameworks of poles covered with bark or animal hides
Resources	rich soil for farming crops of corn, beans, and squash, along with hunting wild game such as deer and birds

Southwest Native Americans	Navajo, Apache, Hopi, Pima, Pueblo, Papago, Cochimi

Spotlight on: Pueblo Tribe	
Region	Four Corners—where Utah, Colorado, Arizona, and New Mexico meet; desert terrain with many cliffs, canyons, and rock formations
Climate	hot and dry with little rain
Shelter	houses built of adobe in cliffs; large apartment-like dwellings for many families with 20 to 1000 rooms
Resources	desert soil good for growing crops such as beans, cotton, and corn; also used to make clay for pottery

Sources: *The World Book Encyclopedia*, 1991 and *Man in Nature*, Carl Sauer, 1980

Historical Native American Shelters

Panoan house

Omaha earth lodge

Navajo hogan

Interior Salish
earth house

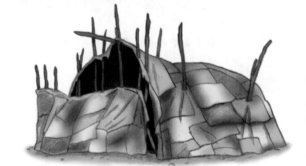

Tehuelche hut

Inca masonry
and thatch hut

Arawak
thatch hut

A Look at Jamestown Through Time

- Jamestown was settled.
- The ground was swampy and the water impure.
- A meager diet weakened the settlers and over half of them soon died of dysentery, malnutrition, malaria, and pneumonia.

- Captain John Smith returned to England for treatment of an injury caused by an accident.
- The settlers had a terrible winter and suffered from hunger.

- The settlers gave up trying to produce silk and grapes.
- They began raising hogs and corn.
- Shipbuilding, glass-blowing, and iron smelting soon began.

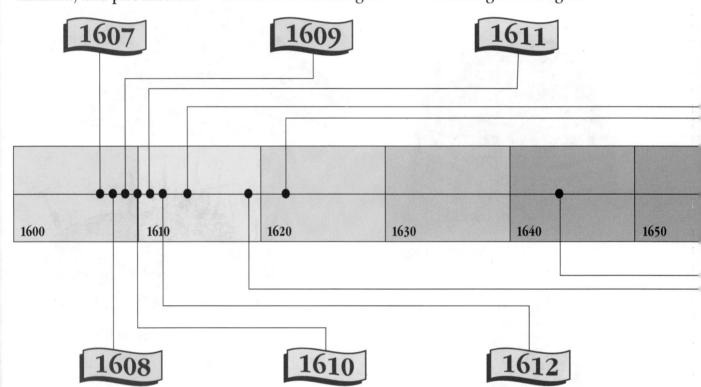

1607   **1609**   **1611**

1600   1610   1620   1630   1640   1650

1608   **1610**   **1612**

- Captain John Smith took control of the settlement.
- He taught the settlers to follow the Native Americans' farming example and urged them to stop looking for gold.

- Governor Thomas West, Lord De La Warr, arrived with supplies and more settlers.

- John Rolfe brought a new type of tobacco from Trinidad. It was a good crop for Jamestown to sell in England.

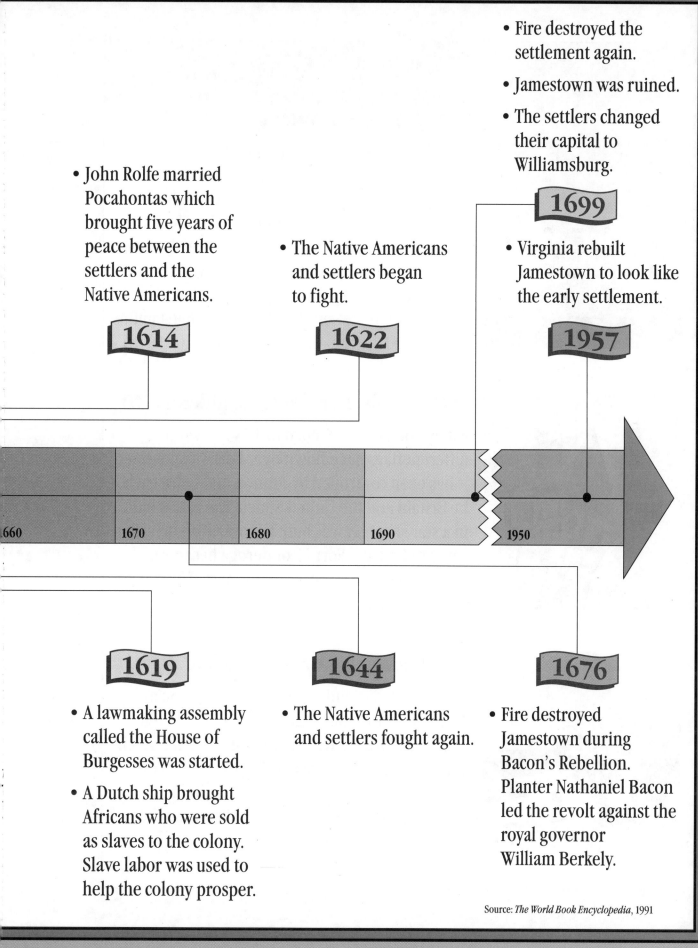

• Fire destroyed the settlement again.

• Jamestown was ruined.

• The settlers changed their capital to Williamsburg.

1699

• John Rolfe married Pocahontas which brought five years of peace between the settlers and the Native Americans.

1614

• The Native Americans and settlers began to fight.

1622

• Virginia rebuilt Jamestown to look like the early settlement.

1957

660 1670 1680 1690 1950

• A lawmaking assembly called the House of Burgesses was started.

• A Dutch ship brought Africans who were sold as slaves to the colony. Slave labor was used to help the colony prosper.

1619

• The Native Americans and settlers fought again.

1644

• Fire destroyed Jamestown during Bacon's Rebellion. Planter Nathaniel Bacon led the revolt against the royal governor William Berkely.

1676

Source: *The World Book Encyclopedia*, 1991

Further Fort Facts

The Alamo • San Antonio, TX

Established as a Catholic mission by the Spanish in 1744, the Alamo consisted of a church and living quarters. It was surrounded by high walls which guarded against enemy attacks. The mission was called the Alamo, the Spanish name for the cottonwood trees surrounding it. It was built of adobe. Adobe is large blocks of mud baked hard in the sun. The Alamo became famous during the Mexican-American War (1846–1848).

Fort Ticonderoga • Fort Ticonderoga, NY

Established by the French in 1755, several different flags have flown over Fort Ticonderoga. It has been controlled by France, and twice each by Britain and the United States. The fort is built in a star shape. It was built to control an invasion route to Canada. Fort Ticonderoga became a stronghold during the Revolutionary War (1775–1781).

Fort Mackinac • Mackinac Island, MI

On a wooded island in Lake Huron high on a bluff, the British built this fort in 1780. A large ramp and fortress were built of limestone, which was plentiful in the area during that time. Fort Mackinac was surrounded by tall timber fences. Many log living quarters were built within the wooden walls.

Isolated Communities of the Past

Buffalo, New York	1803	• The Holland Land Company settled at the head of the Niagara River. The community was called New Amsterdam, but settlers called it Buffalo from Buffalo Creek, a nearby stream.
	1825	• The Erie Canal was built to provide an all-water route to New York City. The canal enabled Buffalo to become an important trade center linking the east and the west.
	1840	• The world's first grain elevator was built in Buffalo. It processed grain shipped from the west over the Great Lakes, then transported the grain to the east on the Erie canal.
	1896	• Large-scale production of electricity began at Niagara Falls.

Salt Lake City, Utah	1847	• The city was founded by 148 Mormon pioneers, led by Brigham Young. Utah was not under United States law. This was important to the Mormons because they did not agree with the United States' rules. • Named for the Great Salt Lake, Salt Lake City is in a valley, surrounded by mountains.
	1850	• Almost 5,000 Mormons lived in the city by this time. • Utah became a United States Territory, and Brigham Young became its first governor.
	1863	• Lead and silver were discovered in Brigham Canyon.
	1869	• Workers laid the tracks that completed the nation's first transcontinental railroad. Salt Lake City could now easily transport its farm and mining products out of the area.
	1896	• Utah became a state, with Salt Lake City as its capital.

Types of Communities

City (or Urban area) an area with 2,500 or more people, usually including a centralized city and nearby smaller communities

Community a place where people live, work, and have fun together

Edge city suburban downtown areas that include shopping malls, office complexes, and industrial parks that are developed near major expressways

Metropolitan area a large city and its inner and outer suburbs

Rural area farms and towns in uncrowded areas with fewer than 2,500 people

Suburb a community near or on the outskirts of a city. Almost half of all Americans live in suburbs.

Town a cluster of houses and other buildings in which people work and live. Usually it is smaller than a city, but larger than a village.

Village any small group of houses and dwellings. A village is often a trade and social center of a township and has a president and a board of trustees.

Source: *The World Book Encylopedia*, 1991

Kinds of Roads — Abbreviations

Kinds of Roads	Abbreviations
Avenue	Ave.
Boulevard	Blvd.
Court	Ct.
Drive	Dr.
Freeway	Fwy.
Highway	Hwy.
Interstate	I
Lane	Ln.
Parkway	Pkwy.
Point	Pt.
Road	Rd.
Square	Sq.
Street	St.

Community Resources

- Library*
- Fire Station
- Post Office
- Police Station
- City Hall
- Chamber of Commerce*

- Grocery Store
- Gas Station
- Hardware Store
- Bank
- Pharmacy

- Video Store
- School
- Church
- College
- Park
- Sports Arena
- Courthouse*

- Barber/Beauty Shop
- Auto Repair Shop
- Shopping Mall
- Hospital
- Restaurant
- Pool

* Resource you can visit or write to for more information about your community

A Mental Map of Place in Space

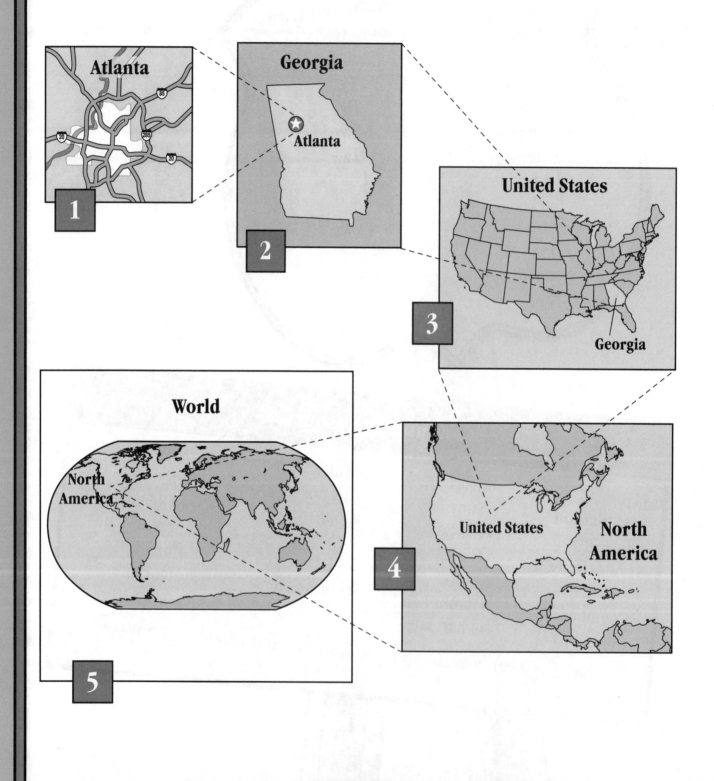

Population Density of the United States, 1995

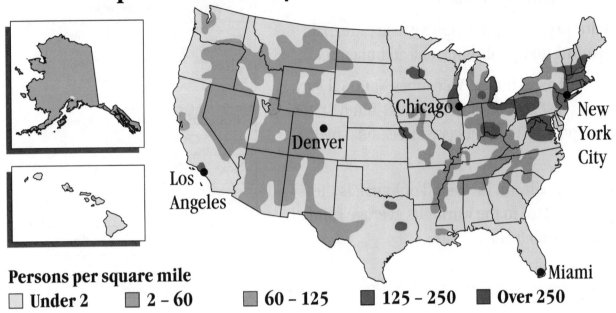

Persons per square mile

⬜ Under 2   ⬜ 2 – 60   ⬜ 60 – 125   ⬛ 125 – 250   ⬛ Over 250

Community Recreational Activities

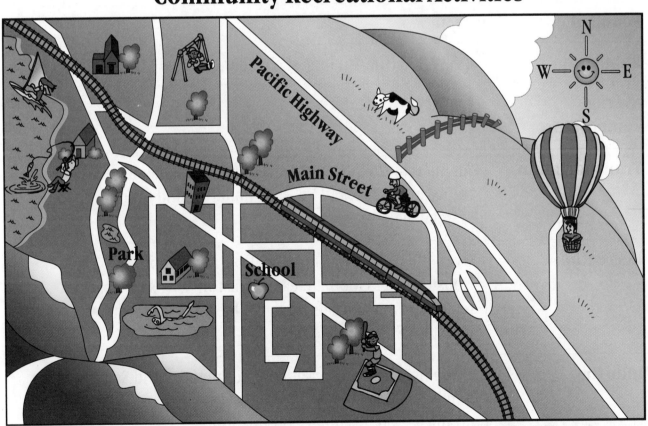

Interesting Facts About Hawaii

- Largest industry: tourism

- Major products: pineapples, sugarcane, fish, printing and publishing, macadamia nuts

- Mauna Loa in Hawaii Volcanoes National Park is the largest active volcano in the United States. It erupts about once every three and a half years.

- Mt. Waialeale, on the island of Kauai, is the rainiest place in the world. Its average rainfall is about 460 inches a year.

Source: *The World Book Encyclopedia*, 1991; *The 1996 World Almanac and Book of Facts*, Funk & Wagnalls

Average Temperature

	Highs	Lows
January	80	66
July	88	74

Population of Three Largest Cities in Hawaii

Rank	City	Population
1	Honolulu (capital city)	377,059
2	Hilo	37,808
3	Kailua Kona	36,818

More About Houston

Founded	1836
Population	1,629,902 (the 4th largest city in the United States)
Average Temperature	January – 55°F July – 83°F
Transportation	An intercontinental and a smaller airport; passenger trains and freight rail lines; an underground tunnel system; major freeways
Recreation	Over 400 parks, Houston Zoo, Houston Symphony, Houston Oilers football team, Houston Astros baseball team, Houston Rockets basketball team, and Houston Livestock Show and Rodeo
Industry	Shipping, oil refining, National Aeronautics and Space Administration (NASA), medical research, textiles, paper products, and machinery production

Source: *The World Book Encyclopedia*, 1991

More About The Four Fabulous Communities

City	Geography	Products	First Settlements and Changes Over Time
Phoenix, Arizona	• Lies in the Salt River Valley, a flat region surrounded by low mountains • Warm, sunny, dry climate	• Railroads made Phoenix a center for trading and processing farm products • Aircraft manufacturing	• Hohokam Native American tribe • United States military post in 1865 • Irrigation canals were used to bring water from the Salt and Gila rivers.
St. Paul, Minnesota	• Located on the banks of the Mississippi River • Dry, cool climate • Fertile land for growing wheat	• Important transportation center due to railroads, airports, and the Mississippi River • Automobiles, transportation equipment, cosmetics, and machinery	• Sioux Native American tribe • Founded in 1840 as an important fur-trading post and a busy river port
Oklahoma City, Oklahoma	• Located near the North Canadian River • Warm, dry climate	• Building materials, cottonseed oil, iron and steel products, paper products • Nation's leading cattle market	• Creek and Seminole Native American tribes • Federal government bought the land in 1889 • Several water reservoirs had to be built
Richmond, Virginia	• Located near the James River • Wet, warm climate • Fertile land for tobacco farming	• Tobacco products, machinery for manufacturing tobacco products, food products, and paper	• Powhatan Native American tribe • Fort Charles built in 1644 to attract settlers • Settlers set up trading posts for furs, hides, and tobacco

Source: *1996 Information Please Almanac*, Otto Johnson, editor

Spotlight on Urban Communities

San Francisco, CA	
Cultural Activities	• War Memorial Opera House • Louise M. Davies Symphony • Herbst Theater (music, dance, and theater performances) • Asian Art Museum • San Francisco Museum of Modern Art • Morrison Planetarium • Steinhart Aquarium
Professional Sports	• San Francisco 49ers—football • San Francisco Giants—baseball
Industry	• Garment making • Food processing • Producing electronic equipment • Printing and publishing • Banking • Tourism • Ports
Transportation	• San Francisco International Airport • Electronic rail system—Bay Area Rapid Transit (BART) • Cable cars, buses, trolleys
Recreational Activities	• Boating and sailing • 160 parks and playgrounds • San Francisco Zoo • Muir National Forest (giant redwood forest)

Chicago, IL	
Cultural Activities	• Chicago Symphony Orchestra • Goodman Theater* • Steppenwolf Theater* • Court in Hyde Park* • Field Museum of Natural History • John G. Shedd Aquarium • Adler Planetarium • The Art Institute of Chicago • DuSable Museum of African American History (*Broadway shows, classics, and comedies)
Professional Sports	• Chicago Bears–football • Chicago Cubs and Chicago White Sox–baseball • Chicago Bulls–basketball
Industry	• Service industry • Clothing • Industrial and atomic research • Ports • Chicago Stock Exchange • Publishing
Transportation	• 3 Airports–O'Hare International (world's busiest airport), Chicago Midway Airport, Meigs Field • Chicago Transit Authority (CTA)–Bus lines, elevated subway, ground level trains
Recreational Activities	• 560 parks and playgrounds • Cook County Forest Preserves

Source: *The World Book Encyclopedia,* 1991

GLOSSARY

adapt: to learn to do something

adobe: a sandy clay mixed with straw used for making bricks

artifacts: old items that tell us about a group of people

artificial: not natural, usually human-made

bayou: a creek or minor river

box canyon: a valley so steep its walls seem almost vertical

business: the occupation, work, or trade in which a person is engaged

cardinal directions: north, south, east, and west

cartographer: someone who draws maps

commute: to travel daily by car, bus, or train to work

compass rose: a map symbol that shows directions

continent: a large body of land made up of many countries

coquina: a soft limestone commonly found in Florida

cultural activities: activities related to art or learning

desert: a large, dry area of land with little or no rainfall

dredging: cleaning, deepening, and widening a waterway

environment: surroundings and features of a place

equator: an imaginary line around Earth's surface halfway between the North and South poles

hill: a high landform, not as tall as a mountain

historical map: a map that gives information about the past

geographer: a person who studies Earth's surface

globe: a model of Earth on which shapes, areas, distances, and directions are all shown

grid: a system of lines that form boxes

gulf: an area of ocean or sea partially surrounded by curved land

indigo: a plant from which purple dye is made

intermediate directions: northwest, southwest, northeast, southeast

island: a piece of land surrounded by water on all sides

isolated: cut off from or not connected to other people or places

lake: a large body of water surrounded by land

lava: melted rock that flows out of volcanoes

legend: a map key that explains the symbols on a map

locator map: a map that helps find where a certain place is

map: a drawing of all or part of Earth's surface on a flat piece of paper

mental map: a way of organizing information in your mind

metropolitan area: a large city and its inner and outer suburbs

mountain: peaked land that rises very high above the land around it

mainland: the one big piece of land that makes up the United States

manufacturing: making or producing a product, especially large-scale and by machines

map scale: a feature used to figure out distance from one place to another on a map

map title: the label that tells what a map is about

natural: made by nature

ocean: the biggest body of water that surrounds land

peninsula: an area of land surrounded by water on three sides

physical map: a map that shows different landforms and water forms

plain: a broad area of low, level land

plateau: a large area of high, flat land

population: all of the people living in a certain area

prairie: land that has many different kinds of grasses growing on it

prime meridian: the starting point to measure both east and west around the globe; located in Greenwich, England

rain shadow: a process in which cool winds from the ocean produce rain that falls in the mountains

resource: something natural or human-made, that people use

river: a large stream of water that flows through land

suburb: a place where people live that is close to or on the outer edge of a city

trade: the business of buying and selling goods

transportation: a system for moving passengers and goods from one place to another

INDEX

1. Study the map. Read each word and its definition.
How many kinds of landforms and water forms are shown on the map? __10__

2. Find the word **mountain** on the physical map.
Is a mountain a landform or a water form? If it is a landform,
write the letter **L** above the word on the map.
If it is a water form, write the letter **W** above the word on the map.

3. Find the word **river** on the physical map.
Is a river a landform or a water form? If it is a landform,
write **L** above the word. If it is a water form, write **W** above the word.

4. Write **L** or **W** above each of the other eight forms of land and water
shown on the map.

5. Using the information you recorded on the map, list each land and
water feature in the correct column below.

Land	Water
desert	gulf
hill	lake
island	ocean
mountain	river
plain	
plateau	

Activity Objective: Learn the names and definitions of different forms of land and water.
Cross-Curricular Connection: Science – Understands basic features of Earth; knows that Earth material consists of solid land and liquid water.
Almanac: See page 110.

Lesson 1

ACTIVITY Create a game to learn forms of land and water.

A Geography Game

The BIG Geographic Question   What are the different forms of land and water?

From the article you learned that land and water are the two
main physical features of Earth. The map skills lesson showed
you some of the different forms of land and water. Now create
a game to help you and your classmates learn the names of
the different forms of land and water.

A. Remind yourself of the main bodies of land and water on Earth.

1. List the seven continents below.

a. Africa   e. Europe

b. Antarctica   f. North America

c. Asia   g. South America

d. Australia

2. List the four oceans below.

a. Atlantic   c. Indian

b. Arctic   d. Pacific

National Geography Standards: 1, 3, 8
6   Basic Geography Skills: Organizing geographic information and answering geographic questions
Geographic Theme: Place – Physical and human characteristics

B. Look back at the different forms of land and water shown on pages 4–5.

1. List the landforms below.

a. desert   d. mountain

b. hill   e. plain

c. island   f. plateau

2. List the water forms below.

a. gulf   c. ocean

b. lake   d. river

C. Write down the names of some of your favorite games. What game could you make to help you learn the names of oceans, continents, landforms, and water forms?

Bingo   Go Fish

Concentration   Jeopardy

D. Think about how you will teach the game to your classmates. Write simple directions for playing your game on the lines below.

(Make sure students' directions clearly explain how to play their games.)

E. Make your game. First, try it out with a family member or friend to see if you need to make any changes. Make the changes, then teach your game to a classmate. (You may wish to invite students to teach their games to each other in small groups or in front of the class.)

---

1. Where is the town square located?

in the center of town or just west of the center of town

2. Write the names of the four streets that surround the town square.

Cherry Street   Granville Street

Columbus Street   Vernon Street

3. Which two of the four streets listed above run north and south?

Columbus and Vernon Streets

4. Which two of the four streets listed above run east and west?

Granville and Cherry Streets

5. List the names of some interesting places in Sunbury and where
they are located.

Place Name	Location
Community Library	west of the town square
Sunbury Water Tower	north of town square, near Sunbury Reservoir
Big Walnut Skate Club	north of Cherry Street, near railroad tracks
Big Walnut Elementary School	on Columbus Street, south of the town square
Big Walnut Creek	runs north and south along the east side of town

6. Now cover up the map of Sunbury with a piece of paper. Try to
remember what it looked like. Draw a picture of it in your mind.
Using your mental map and your notes from questions 3, 4, and 5
above, draw a map of Sunbury on paper. When you are finished,
uncover the map in your book and compare it to the one you drew.
How accurate was your mental map?
(Check students' maps drawn from memory for similarities to the Sunbury map.)

Activity Objective: Determine one's own place in space.
Cross-Curricular Connection: Language Arts – Effectively gathers/uses information for research; asks and seeks to answer questions regarding the
characteristics of people and places in one's local community.
Almanac: See pages 112–113, 116–117, and 128.

Lesson 2

ACTIVITY Make a mental map of your own "place in space."

Your Place, Your Space

The BIG Geographic Question   Where do you live?

In the article you learned about the many places that you can say
you live. The map skills lesson gave you some practice in making
a mental map. Now you will make a mental map of your place in
space to show the many places you live.

A. A mental map can be of a big or small space. It can be specific or general.

1. Write the name of your community or town.

(Students should provide answers specific to their place.)

2. List some of the land and water features, buildings, and fun places
to go in your community. (Make sure the places and features students list
are in or likely to be in their communities.)

Land Features	Water Features	Buildings	Fun Places to Go

---

3. Think about the features you listed in the chart and what your
community or town looks like.

4. Draw a mental map of your community or town like you did for
Sunbury.

**B. Think about the bigger, more general areas that include
your community or town. To help you with this, answer the
following questions.** (Check students' answers and mental maps to see whether they show an
understanding of the various places in space that they occupy—home,
community, city, state, U.S., North America, and world—in terms of scale.)

1. Near what bigger city is your community or town located?

2. In what state is your community or town located?

3. Use the United States map in the Almanac to locate your state.

4. Use the world map in the Almanac to locate the United States.

**C. The "bull's-eye" mental map that you saw on page 9 is one way
to show your place in space. Look in the Almanac for an example
of a mental map for a person who lives in Atlanta. Now create a
mental map that shows your place in the bigger space of the world.**

A. Look at Maps A and B.

1. Which map tells you which bus to take to Library Square?

   Map B

2. Which map can help visitors find their way around an elementary school?

   Map A

B. Answer these questions about Map A.

1. What is the title of Map A? Sturgeon Elementary School, 1975

2. Does the school have a gym? Yes

3. Draw a line to show how to go from the library to the supply room.
   (Students should draw a direct route using hallways from library to the supply room.)

4. Write one sentence describing the information this map gives.

   It shows where different rooms are located in the Sturgeon Elementary School.

5. What do you think this map can be used for?

   It could be used by new students and visitors to find their way around.

C. Answer these questions about Map B.

1. What is the title of Map B? DASH Downtown Los Angeles Rapid Transit

2. Which bus route takes you to Library Square? Route E

3. Where does the Metro Red Line take you?

   Union Station or downtown from Union Station

4. Write one sentence describing the information this map gives.

   It shows how to get to different places by bus or Metro Red Line in downtown Los Angeles.

5. What do you think this map can be used for?

   It could be used to help tourists find their way around the city by bus or Metro Red Line.

Activity Objective: Make a map that effectively provides information and communicates a message.
Cross-Curricular Connection: Language Arts — Effectively gathers/uses information for research; presents information obtained from research in a display that integrates multiple representations of information
Almanac: See page 129.

Lesson 3

ACTIVITY
Make a map that provides information and communicates a message.

Mapping a Message

The BIG Geographic Question — How do maps communicate information?

From the article you learned some of the basic features of a map and their purposes. The map skills lesson showed you that different kinds of maps are used to communicate different kinds of information. Now create a map that communicates information to a family member or friend.

A. Read the below list of different types of maps. Circle the type of map that you think you would like to create. You may want to look in the Almanac for examples of other types of maps, such as a recreation or population map. Or you can add your own ideas to the list below. (Make sure students select a type of map to create.)

a map of how to get to your room in your house

a neighborhood street map

a map of your school bus route

a map of the land and water features of a nearby park

a map of where you hid something last year

_____

_____

_____

Informal Assessment: Group students according to the types of maps they made. Have students tell how their maps are alike and different. Help them see that even maps about a similar topic can be different in many ways. Have them discuss why some maps are easier to understand than others.
Standardized Test-Format Assessment: See page 149.

B. Make a list of things to include on your map. Look back at the article on pages 14 and 15 for suggestions. (Make sure students list key map elements.)

_____   _____

_____   _____

_____   _____

C. Think about your map's legend. Choose four things on your map that you could show as symbols. Write one of the names in the space at the top of each box. Then draw the symbols in the larger boxes. (Make sure students' symbols and names match and are clear.)

D. Make a practice sketch of your map so you can plan where to put everything. (Check students' sketches for necessary map elements.)

E. Make a final version of your map. Have a family member, friend, or classmate read your map and describe the information or message shown. Was their answer correct? Did your map communicate the message you wanted? If not, how could you change it to make it more clear? (You may wish to have students compare their maps.)

A. If you were at Petersburg National Battlefield, which direction would you go to get to these places?

1. medical center west

2. post office south

3. visitors center north; northwest

Northeast is an **intermediate direction**–a direction halfway between two cardinal directions. The other intermediate directions are northwest, southeast, and southwest.

4. In which direction would you travel from the battlefield to the

   Appomattox River? northwest

5. In which direction would you go from the Medical Center to

   Southpark Mall? southeast

B. Follow these directions. Where are you?

1. Start at the high school. Go north on Johnson Rd. to Sycamore St. Go south on Sycamore St. to South Blvd.

   the elementary school

2. Start at the post office. Go north on Crater Rd. to Washington St. Go west on Washington St. to Sycamore St. Go south on Sycamore St. Stop at the corner of Marshall St.

   the library

C. Now that you can find your way around Petersburg, you can give directions to someone else.

1. Tell how to get from the high school to Poplar Park.

   north on Johnson Rd., north on Sycamore St., east on Fillmore St.

2. Tell how to get from the Courthouse to Southpark Mall.

   south on Sycamore St., south on Crater Rd., east on South Park Dr.

Activity Objective: Find a place in one's community and map a route to get to a place in the community.
Cross-Curricular Connection: Language Arts — Demonstrates competence in the general skills/strategies of the writing process.
Almanac: See page 127.

Lesson 4

ACTIVITY
Use a map to show how to find a place in your community.

Where Are You?

The BIG Geographic Question — How can you help visitors find their way around your community?

In the article you read about some of the people and places in the community of Petersburg. The map skills lesson helped you find your way to some of the important places. Now use what you know about directions to make a map of your community.

A. Make a list of important places in your community. An example has been done for you. (Make sure the places students list are in their community.)

1. school

2. _____

3. _____

4. _____

B. Think of symbols to stand for the places on your map. Draw a symbol in each box. (Map key symbols should be appropriate.)

1.	2.

National Geography Standards: 1, 2, 3, 4
Basic Geography Skills: Asking geographic questions and organizing geographic information
Geographic Themes: Location – Position on Earth's surface; Place – Physical and human characteristics

---

Informal Assessment: Have students who have given directions to the same place compare their maps and directions. Encourage them to help one another clarify their directions.
Standardized Test-Format Assessment: See page 150.

3.	4.

C. List at least five streets that will be on your map.

1. (Students should list streets near the places they named in A.)

2. _____

3. _____

4. _____

5. _____

D. Draw a map of your community. Be sure your map includes all the places and streets from your lists. Also include a map key. (Check to see whether students' maps accurately indicate places in the community.)

E. Write directions.

1. Choose one of the places on your map. Draw the way to go from your school to that place. (Make sure students have drawn a route on their maps.)

2. Write directions from your school to that place. Be sure to include street names and compass directions. (Make sure students' directions are accurate according to their maps.)

_____

_____

_____

A. Use the map index and grid to find these places.

1. Lincoln Memorial F3
2. FBI Building C12
3. Smithsonian Institution F11

B. Use the map grid to identify these places.

1. What important home is found at B8? White House
2. A memorial to the third U.S. president is found at I7. What is its name?
   Jefferson Memorial
3. A major transportation center is located at A19. What is it?
   Union Station

C. Use the map index and grid to answer these questions about the Mall.

1. What is at the east end of the Mall? Capitol building
2. What is at the west end? Washington Monument
3. Identify four of the important buildings found along the Mall.
   (Answers might include: Smithsonian Institution, Museum of Natural History, National
   Gallery of Art, National Air and Space Museum, and Museum of American History.)

Activity Objective: Compare a planned and an unplanned community.
Cross-Curricular Connection: History – Understands the history of the local community; understands changes in land use and economic activities in the local community since its founding
Almanac: See page 126.

Lesson 5

ACTIVITY Compare your community's history with the history of Washington, D.C.

Your Community: Was It Planned?

The BIG Geographic Question How can you tell whether you live in a planned community?

From the article you learned how the community of Washington, D.C., was planned. The map skills lesson showed you how to use a grid system. Now find out about the history of your own community to figure out whether it was planned.

A. Answer these questions to find out about the history of your community. (Answers to these questions should reflect information students find about the early history of their community.)

1. Who were the first people to settle in your community? _____

2. When was the community started? _____

3. Why was the community started? _____

B. Complete the chart below to compare facts about your community to Washington, D.C.

	History of Washington, D.C.	History of Your Community
Why was that location selected?	central location; close to water	(Students' answers should reflect information students find out about their community.)
What were its major physical features?	Potomac River	
What is the purpose of the community?	the capital of the new nation	
Were streets planned using a grid?	yes	
Were parks included in the plan?	yes	
Where are many important buildings located?	near or around the mall	

C. Think about how your community grew and complete the following. (Students might work together on the maps of their community.)

1. Find or make a map of your community as it is today.
2. Tell how your community has grown. (Encourage students to share sources of information about the growth of their community and then compare conclusions.)
   _____
   _____

3. Do you think your community was planned?
   _____
   _____

Materials: index cards; information about your local community (such as physical features, important buildings, and so on)

A. Different maps use different symbols to show the capital city of a state or country. You can find each symbol in the map key. Look at the map of the United States. Find the map key.

1. What is the symbol for the United States capital? ✪
2. What is the symbol for the state capitals? ★

B. Find the states of Arizona, Minnesota, Oklahoma, and Virginia on the map. Use the information on the map to find each state's capital and the community described in the article for each state. Record your information on the chart below.

State	Capital City	Community Featured in Article
Arizona	Phoenix	Oraibi
Minnesota	St. Paul	Leech Lake
Oklahoma	Oklahoma City	Oklahoma City
Virginia	Richmond	Abingdon

C. Choose one of the capital cities you listed above to learn more about. Look in the Almanac for information on the state's capital. Complete the T-chart below by listing information about the capital city and the featured city. Note any similarities and differences. Based on the information you have collected, decide whether the state's capital city is representative of other unique communities in the state. (An example follows.)

Capital City	Featured City
Phoenix, Arizona • located in the Salt River Valley • a flat region with low mountains • warm, dry climate • large vacation industry • Hohokam Native Americans lived there	Oraibi, Arizona • located in northern Arizona • plateaus, or high, flat land • strong winds shape the land • steep canyons of red rock • Hopi Native Americans lived there

---

Activity Objective: Recognize the geographical features of one's own community that make it a special and attractive place to live and visit.
Cross-Curricular Connection: Language Arts – Effectively gathers/uses information by asking and seeking answers to questions about people/places in the local community
Almanac: See page 131.

Lesson 6

ACTIVITY Advertise physical features of your community that attract people to live or visit there.

Your Community: What Are Its Features?

The BIG Geographic Question What physical features of a community make it a unique place?

In the article you read about some of the physical features that make four communities special. In the map skills lesson you located state capitals and looked at whether they were representative of other communities in the state. Now plan and present an advertisement that highlights a physical feature of your community to attract people to come there.

A. Think about the features of the communities you read about on pages 32–33.

1. Use the chart below to list the important physical features of each community. An example has been done for you.

State	Arizona	Minnesota	Oklahoma	Virginia
Community	Oraibi	Leech Lake	Oklahoma City	Abingdon
Features	• sandy red plateaus • wind and rivers carve canyons through red rock	lakes, rivers, forests	plains	mountains, forests

---

Informal Assessment: Help students assess the advertisements based on special features highlighted, design appeal, and whether they might attract people to their community.
Standardized Test-Format Assessment: See page 152.

2. How do these features help make the communities special places to live or visit?

Oraibi: majestic landscape, Hopi Native American community, beautiful pottery; Leech Lake:

wild rice, Ojibway community lifestyles, beautiful forests; Oklahoma City: Native American community

and celebrations; Abingdon: mountain towns, forests, traditional crafts

B. Think about your community and complete the following.

1. Make a chart showing the special physical features of your own state, its capital city, or your community. Some of these physical features might include rivers, lakes, forests, mountains, or plains. Some other features might include the different groups of people who live there or fun things to do there. Jot down your notes below. (Students' answers should accurately reflect their communities.)

Your Community's Name	
Special Physical Features	
Other Special Features	

2. How do these features make your community special?

(Students' answers should accurately reflect their communities.)

C. Now plan and present an advertisement for your community. Your advertisement can be a poster, a flyer, a TV commercial, or a radio commercial. It might include a picture and a written description of the physical feature that makes your community a unique place to live or visit.

(Make sure students' ads include a description of the physical feature(s) that make their community special. You may wish to invite students to share their commercials with the class.)

Materials: information about your local community

LESSON 7

A. Look at the map of early Virginia.
1. Circle Jamestown.
2. Write down the time in history that this map is showing. 1607–1612
3. Circle the Chesapeake Bay and Atlantic Ocean.
4. Mark an X on the early Native American settlements.
   (Students should mark Susquesahannock, Powhatan, and Chickahominy.)

B. Look at the map and complete the following.
1. Circle the names of the bodies of water you see on the map.
2. Write the names of the bodies of water.

   a. Atlantic Ocean   c. James River
   b. Chesapeake Bay   d. Chickahominy River

3. Trace the peninsula on which Jamestown is located. (Make sure students correctly trace the peninsula.)
4. Write the names of the bodies of water Captain Smith followed to get from Jamestown to:

   a. Susquesahannock (sahs-kwah-sah-HAN-ak)
   James River, Chesapeake Bay, James River

   b. Powhatan (pow-HAT-ahn)
   James River

   c. Chickahominy (chi-ka-HĂ-ma-nē)
   James River, Chickahominy River

5. Was the peninsula a good location to settle Jamestown? Why or why not?
   (If students said yes, the reason might be: Water provided a way to travel and move goods from place to place. If students said no, the reason might be: The water was not safe to drink.)

6. Why was it important to be located near water in the early days of the Jamestown settlement?
   Waterways were used for transportation, exploration, and trade.

LESSON 7

Activity Objective: Consider the reasons for staying in or leaving an early settlement, and make a decision.
Cross-Curricular Connection: History – Understands how communities in North America varied long ago.
Almanac: See pages 122–123.

Lesson 7

ACTIVITY
Decide whether you would have left or stayed in Jamestown.

What Would YOU Do?

The BIG Geographic Question
Would you have remained in Jamestown or gone back to England in 1607?

A. List on the idea web below characteristics of life in early Jamestown. Include both positive and negative characteristics. Add more spaces if you need to. (Characteristics of Jamestown might include the following.)

- swampy land
- attacks by Native Americans
- disease
- hard work
- Characteristics of Jamestown
- bad drinking water
- more freedoms
- hopes of getting rich
- possible jobs

National Geography Standards: 4, 12, 17
Basic Geography Skills: Acquiring, analyzing, and organizing geographic information
Geographic Theme: Movement – Humans interacting on Earth

42

LESSON 7

Informal Assessment: Have students compare their reasons and decisions for staying or leaving. Discuss them as a class.
Standardized Test-Format Assessment: See page 153.

B. Use the chart below to take notes on what the living conditions were like in Jamestown in 1607.

	Good	Bad
Land Features	many trees	swampy land
Water Features	many rivers for travel and transport	bad drinking water
Other Features	more freedoms, hopes of getting rich, adventure, possible jobs	disease, attack by Native Americans

C. Using the information you collected, list reasons you would leave or stay in Jamestown. (Students' reasons might include the following.)

Reasons to Leave	Reasons to Stay
not enough food	promise of riches
bad drinking water	adventure in a new land
wet and swampy ground	many more freedoms
hard work	possible jobs
harsh weather	trees for building

D. Pretend to be an early Jamestown settler. With your classmates, stage a town meeting to discuss whether to leave or stay in the community. Listen carefully to the ideas of others. When the meeting is over, think about the arguments you heard both for staying and for leaving. What did you decide? Write your decision and the reasons for it on the lines below.

(Make sure students' decisions and reasons are appropriate to the time in history.)

43

LESSON 8

A. Look at the map key. Put the appropriate number on the map to show where each of the following physical features is located. An example has been done for you. (Make sure students put numbers in the right places.)

1. mountain   3. lake   5. Great Plains
2. river   4. forest   6. ocean

B. Look at the map and map key to regional groups.

1. Which color shows where the Iroquois tribe lived? red

2. The southwestern United States has many desert areas. Which Native Americans of long ago lived in this area? (Possible answers include the Hopi and the Apache.)

3. Which groups of Native Americans might have used resources from the ocean? (Possible answers include the Chinook, the Maka, and the Delaware.)

4. On what river might Native Americans from the Great Plains region have traveled north and south? They might have traveled on the Mississippi River.

5. Which groups obtained their resources from forests?
(Possible answers include the Iroquois and the Chinook.)

6. What three regions were divided by mountains?
The Great Plains, the Southwest, and Great Basin and Plateau.

7. How do you know that the Chickasaw and the Cherokee came from the southeast?
They are shown to have lived in the green area on the map. Green stands for southeast.

8. If you traveled by land from your state to visit a Native American tribe of long ago in the northwest, what landforms or water forms would you have had to cross? If you lived in the northwest, think about traveling to visit a tribe in the southeast. Explain whether your trip would be difficult.

(Students' answers should reflect an understanding of how to read a map for the location of various landforms such as mountains, rivers, and plains.)

47

LESSON 8

Activity Objective: Use geographical characteristics of a regional area and the tools and artifacts of different Native American tribes as clues about the areas in which they lived and their lifestyles.
Cross-Curricular Connection: History – Understands how to analyze chronological relationships and patterns; knows how to interpret time lines
Almanac: See pages 118–121.

Lesson 8

ACTIVITY
Find out about the ways of living of Native American tribes long ago.

Digging for Clues

The BIG Geographic Question
What can you find out about the first Native Americans by studying the artifacts they left behind?

From the article you learned about the tools, crafts, clothing, and shelters of some Native American tribes of long ago. The map skills lesson showed you where they lived and some of the landforms found in those regions. Use these clues and information you find in the Almanac to learn about the way of living of a Native American tribe of long ago.

A. Complete the following.

1. List the names of Native American tribes you have read about.

Apache	Kickapoo
Cherokee	Miami
Chinook	Navajo
Hopi	Pueblo
Iroquois	Sioux

2. Write down the name of a Native American tribe that you would like to learn more about.

(Make sure students choose a tribe.)

LESSON 8

B. Answer the following questions about the tribe you have chosen.

1. In what region of the United States did the tribe you selected live?
(Make sure students indicate the correct region for the tribe they selected.)

2. What were some of the landforms or water forms found where this tribe lived?
(Landforms should include those shown on the map, such as mountains, forests, plains, and deserts, and water forms such as rivers and lakes.)

3. What was the climate of their land like? (Make sure students indicate the correct climate for their chosen region.)

4. What kinds of resources were found there? (Resources could include water, land, animals, trees, soil, and rocks.)

5. How did the tribe obtain and use these resources? (Possible answers include hunting, carving, digging, chopping, or gathering.)

C. Make some additional notes about the tribe that you have chosen using the article and Almanac. Make a chart like the one below and record your information on it. (Sample information is provided.)

Native American Tribe Name and Region Iroquois of the northeast region

Shelter	Clothing	Food	Tools	Crafts	Movement
• wooden poles covered with bark	• wampum belts • animal skins	• beaver • deer • bear • rabbit • muskrat • corn	• stone axes • wooden tools	• masks • stone bowls • baskets • cornhusk dolls	• canoes made of birch bark • walking with moccasins or snowshoes

D. Use the clues that you gathered to tell a story about the life of the Native American tribe that you studied. Design a poster that shows what you discovered. Include details about how the people worked and lived. Plan your poster first, then make it. Include it as part of a classroom display about early life in America.

(Students' displays might include illustrations, diagrams, and captions. Dioramas could be constructed from materials such as shoe boxes, wood, bark, leather, paper, string, and clay.)

49

**A. Look at the map on page 72 of the United States in the 1700s.
Circle the following:** (Make sure students have circled the correct items on the map.)

1. map key
2. St. Augustine
3. Boston
4. rice symbols
5. cattle symbols
6. Atlantic Ocean

B. Look at the symbols on the map key and answer the following.

1. Write down what each symbol represents.

   a. 🌲 is the symbol for _lumber_

   b. 🐟 is the symbol for _fish_

   c. 🌿 is the symbol for _tobacco_

2. Find Charleston on the map. In what early colony is Charleston located?

   South Carolina

3. Look at the symbols on the map near Charleston. What resources can be found in or near Charleston?

   cattle _____ indigo

4. What towns have indigo as an important resource?

   Savannah _____ Charleston _____ St. Augustine

5. What town has coquina stone as an important resource?

   St. Augustine

6. Which resource is found the most often on the map?

   cattle

7. What could the people of North Carolina have done with their resources?

   (Students' answers might include: build homes and ships with the lumber; make cloth from the

   cotton; and eat and export the rice and grain.)

_____

53

Activity Objective: Compare resources of communities of the past to those of today.
Cross-Curricular Connection: History – Understands changes in land use and economic activities in the local community since its founding
Almanac: See page 124.

Lesson 9

ACTIVITY Find out what resources are found in your community.

Your Community's Resources

The BIG Geographic Question — How have local resources been used by your community?

From the article you learned how the town of St. Augustine was started and that its resources were used to build forts and homes. The map skills lesson showed you where different resources were found in different places in the United States long ago. Find out what resources are located in your community. Remember, people are resources, too.

A. Answer the following questions about your community.

1. What farm crops are grown in your community? _____

   (Make sure the crops students list are grown in their community.)

2. What farm animals are raised in your community? _____

   (Make sure the animals students listed are raised in their community.)

3. How do some of the people in your community earn a living? _____

   (Make sure students list occupations that exist in their community.)

4. Write *yes* in the blank if the following are found in or near your community.

   lake, river, ocean _____

   woods _____

   quarries or mines _____

   stores, hospitals, schools _____

54

National Geography Standards: 4, 14, 16
Basic Geography Skills: Asking and answering geographic questions and organizing and analyzing geographic information
Geographic Theme: Place – Physical and human characteristics

B. Write four resources found in your community. Then write how each resource is used. (Students should list four resources found in their community and possible uses of each.)

Resource: _____ Use: _____

Resource: _____ Use: _____

Resource: _____ Use: _____

Resource: _____ Use: _____

C. Using the chart below, compare your community's resources with those of St. Augustine. First write the four resources you listed above. Then put a check mark in each box if your community or St. Augustine has that resource. (Make sure students have accurately completed the chart.)

Resource	Your Community	St. Augustine
	✓	
	✓	
	✓	
	✓	
stone		✓
lumber		✓
water (nearby ocean, lake, or river)		✓

D. Talk to friends and family members or go to the library to find out whether your answers were correct. Use or draw a map of your community showing where its important resources are located. Write a short description of why these resources are important to your community.

(Students' maps should reflect knowledge of their community. Answers should reveal an

understanding of the importance of local resources to a community.)

Materials: information about your local community (including resources such as crops, animals, natural resources, and people)

55

A. Cover up the map key on page 58. Now look at the symbols below. Write what each symbol stands for.

🐄	cattle	🌼	flowers
🌾	wheat	🥛	milk
🌽	corn	🐔	chicken, poultry
🍎	fruit	🍃	tobacco
🐟	fish	🌸	cotton

B. Write the name of a product that comes from each region listed below. Then write the states where each product is found.

Region	Product	State
New England	fish	ME, VT, MA, NH
Mid-Atlantic	milk, fruit	NY, PA, NJ
Southeast	poultry, cotton, fruit	AR, LA, MS, AL, GA, SC, FL
Heartland	cattle, corn, wheat	ND, SD, NE, KS, MN, IA, MO
Southwest	cattle	TX, NM, AZ
Pacific	fish, fruit	WA, CA, OR
Alaska	fish	AK
Hawaii	fish, fruit, flowers	HI

C. List the regions that produce the most:

cattle _Heartland, Mountain, Southwest_

corn _Great Lakes, Heartland_

tobacco _Appalachian Highland_

Activity Objective: Plan an event to exhibit the things that rural communities are known for and proud of.
Cross-Curricular Connection: Language Arts – Effectively gathers/uses information for research purposes about people and places within (or outside of) the local community
Almanac: See pages 111 and 126.

Lesson 10

ACTIVITY Plan a fair to show the products a rural community might be known for and proud of.

Come to the Fair

The BIG Geographic Question — How is a fair or festival in a rural community an example of people interacting with the environment?

From the article you learned of some rural communities that hold fairs and festivals. The map skills lesson showed you products people grow on farmlands or catch in nearby waters. Now plan a fair to display a product of which a rural community is proud.

A. Use information from the article, the map, and the Almanac to help you choose a place and a product.

1. Choose a small town. _pages 56–57 or one they have visited or heard about._ (Students might choose a location that was highlighted in the article on

2. Describe the location of the town. (Students should give the geographic location

   of the town they selected.)

3. Write the name of a product that is important to the town.

   (Students should use the map or photographs to name a product important to the community.)

4. Tell whether the product grows on the land or comes from the water.

   (Students' answers might include: crops, animals, flowers, and trees grow on land; fish live in water.)

B. Explain why you chose the place and product.

   (Possible answers might include liking the product or knowing someone who lives in the region it

   comes from.)

60

National Geography Standards: 6, 16
Basic Geography Skills: Asking and answering geographic questions and acquiring and analyzing geographic information
Geographic Theme: Human/Environmental Interaction – Relationship within places

---

Informal Assessment: After students complete section C, observe whether they need additional explanation of the suggested events. Take a field trip to a local festival or ask students who have attended one to share their experiences.
Standardized Test-Format Assessment: See page 156.

C. Read the events on the checklist below. Decide which ones you will have at your fair and write *yes* or *no* in that column on the chart. Describe the events you will have at the fair. Remember to feature the local product. (The description of each event should reflect the featured product.)

Event	Yes or No	Description
parade		
displays		
floats		
food		
games		
posters		
music		
contests		
craft booths		
sales		
speakers		
the mayor		

D. Make a poster to advertise your festival. Include these details.

- festival name (The festival name, place, events, and product should be on the poster.
- place   Note whether the poster shows an awareness of the growing season or
- date   harvest time.)
- events
- artwork

Lesson 11
MAP SKILLS
Using a Map and Chart to Find Out Where People Live

Mountains, lakes, and rivers are some of the physical features that can be represented on a map. Population is one of the human features.

A. Look at this map of the United States. Write the names of the following physical features that are shown.

1. Rivers ___Mississippi, Missouri, Ohio, Columbia, Colorado, Rio Grande___

2. Mountains ___Appalachian Mountains, Rocky Mountains, Cascade Range, Sierra Nevada Mountains___

3. Lakes ___The Great Lakes (Huron, Ontario, Michigan, Erie, Superior); Great Salt Lake___

4. Oceans ___Atlantic Ocean, Pacific Ocean___

B. Read the city names and their populations on the chart.

1. Complete the chart by ranking the cities from 1 to 10 from largest to smallest population.
2. Locate each city on the map.

Population of Large Cities in the United States

City	Population*	Rank
Chicago, IL	2,768,483	3
Dallas, TX	1,022,497	7
Detroit, MI	1,012,110	9
Houston, TX	1,690,180	4
Los Angeles, CA	3,489,779	2
New York, NY	7,311,966	1
Philadelphia, PA	1,552,572	5
Phoenix, AZ	1,012,230	8
San Antonio, TX	966,437	10
San Diego, CA	1,148,851	6

*Population as of July 1, 1992

C. List the cities on the map that are located near each of the following: (Possible answers include the following.)

Rivers ___Memphis, New Orleans, Omaha, Kansas City, St. Louis, Minneapolis, Cincinnati___

Lakes ___Milwaukee, Chicago, Buffalo, Cleveland, Salt Lake City___

Oceans ___Boston, New York, San Francisco, San Diego, Seattle, Miami, Baltimore, Los Angeles___

D. Why do you think so many large cities are located on or near water?
Being near water makes travel and trade easier; products can be easily transported by water; water was a power source.

Activity Objective: Explore an important aspect of an urban community.
Cross-Curricular Connection: Language Arts – Effectively gathers/uses information for research purposes; asks and seeks to answer questions about people/places within or outside of the local community
Almanac: See pages 126 and 132–133.

Lesson 11
ACTIVITY
Explore an important feature of an urban community.

Investigating a City

The **BIG** Geographic Question — What makes an urban community a city?

From the article you learned how one city grew to become a big urban area with increased business, transportation, and population. In the map skills lesson you looked at the United States' largest cities that are located near water. Now investigate the human features of your own or a nearby urban community.

A. Listed here are five important features of a city. Describe what each one means. Use the Glossary or Almanac in the back of your book to help you.

1. Population ___All of the people living in a certain area.___

2. Transportation ___A system for moving passengers and goods from one place to another.___

3. Business ___The occupation, work, or trade in which a person is engaged.___

4. Manufacturing ___Making or producing a product, especially large-scale and by machines.___

5. Cultural activities ___Activities people do to celebrate art, education, or people from different cultures.___

National Geography Standards: 9, 10, 11, 12
Basic Geography Skills: Asking and answering geographic questions and analyzing geographic information
Geographic Theme: Human/Environmental Interaction – Relationship within places; Place – Physical and human characteristics

---

B. Complete the following to find out more about important features of an urban community. (Students' answer will vary according to the urban community they select.)

1. Write the name of the urban community where you live or the one closest to you. (Students' answer will vary according to the urban community they select.)
2. Estimate the population of the city. (Students' answer will vary according to the urban community they select.)
3. Fill in this chart with examples you think describe the city you chose. (Students' answer will vary according to the urban community they select.)

Features	Examples
Population	
Transportation	
Business	
Manufacturing	
Cultural Activities	

4. Check with a family member, librarian, or encyclopedia to see whether the examples you listed above are correct.

C. Choose one of the features of the city you selected to explore further. Jot down some notes about the feature.

(Students' information will vary depending on the feature they choose to explore.)

D. Design a banner that celebrates and tells something about the community you studied. The banner design should emphasize the feature on which you focused.

(Make sure students' banners include information about the population, housing, stores, transportation routes, businesses, or places for cultural and recreational activities of the city they chose.)

Map Skills Objective: Use a map to understand geographical terms.
Cross-Curricular Connection: History – Knows how to identify patterns of change and continuity in the history of the community

Lesson 12
MAP SKILLS
Using a Map to Understand Geographical Terms

A **metropolitan area** is a large city and its inner and outer suburbs. The city must have a population of 50,000 or more. The inner suburbs are the residential areas closest to a city. The outer suburbs are residential areas that develop as highways extend farther out.

A. Look at the metropolitan area map of Phoenix.

1. Color the city red.
2. Color the inner suburbs blue.
3. Color the outer suburbs green.

(Make sure students have colored the three parts of the map according to the directions.)

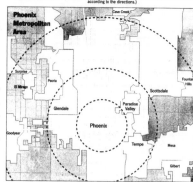

B. Complete the following about the area surrounding where you live. (Make sure students correctly identify the area in which they live.)

1. Write the name of the town in which you live. _____

2. Write the names of other suburban areas near you. _____

3. Write the name of the largest city near you. _____

C. Using the map of Phoenix as a model, draw the area in which you live. Include your town, other cities around your town, and the nearest large city. (Students should sketch a map showing the city in relation to where they live.)

Activity Objective: Explore how a nearby suburban community developed.
Cross-Curricular Connection: Language Arts – Effectively gathers/uses information for research purposes; asks and seeks to answer questions about the local community.
Almanac: See page 135.

Lesson 12

ACTIVITY Explore how a nearby suburban community developed.

Exploring a Suburb

The BIG Geographic Question As a suburb develops, how does the area change?

From the article you learned that people leave both the city and the country to live in the suburbs. The map skills lesson helped you look at the area in which you live in relation to suburbs. Now find out how a suburb near you has grown.

A. Write about the area where you live. (Make sure students are able to describe their community as a city, a rural town, or a suburban town; name surrounding cities and towns; and describe specific means of transportation.)

1. Name the area where you live. _____

2. Is it a city, rural town, or suburb? _____

3. What is a nearby major city? _____

4. What are some nearby suburbs? _____

5. How do people travel between the suburbs and the city or between two suburbs? _____

6. Where do your family members work? _____

7. If they work outside the home, how do they travel to work? _____

8. Where do you go to school? _____

9. How do you get to school? _____

10. Where do you shop? _____

11. Where are the stores, and how do you get to them? _____

B. Think about the houses, roads, open land, stores, schools, and transportation in your suburb or a nearby suburban community. Are they new? Have they always been there? How have they changed? Write some questions you can ask someone to help you find out.

Features	Questions to Ask
Houses	(As students think about the features in their own or a nearby suburban community, they should write specific questions they will ask an adult.)
Roads	
Open Land	
Stores	
Schools	
Transportation	

C. Now interview an older family member, a neighbor, or a school employee. Ask your questions. Take notes on what you learn.
(Make sure students take notes as they conduct their interviews.)

_____
_____
_____
_____
_____
_____

D. Decide how you will share your information. Then present what you learned about suburbs in your area. (Students should make an oral or written presentation along with a recording, a graphic, or a visual display.)

_____ report _____ picture time line

_____ photographs, drawings _____ tape recording

B. Study the map and answer the following.

1. Find the city of Langdon, North Dakota. Grain is harvested on a farm, then loaded onto a truck and taken to Langdon.

   a. What direction will the truck travel to get from Langdon to Highway 2? south

   b. What road will the truck driver take to get from Langdon to Highway 2? Highway 1

2. The grain is stored in Duluth, Minnesota.

   a. What direction will the truck travel to get to Duluth? east or southeast

   b. Trace Highway 2 to Duluth. Why should the truck driver allow extra time? Part of Highway 2 is under construction.

3. The grain is sold to a mill in Minneapolis, Minnesota. There it will be ground into flour.

   a. What direction will the truck travel to get from Duluth to Minneapolis, Minnesota? south

   b. What road will the truck driver use? Highway 35

   c. What does the special symbol around the highway number mean? It means the highway is an interstate.

   d. How would you give directions from Duluth to Minneapolis? Go south on Interstate 35.

4. The flour is sold to a bakery in Sioux City, Iowa. There it will be made into bread. Write directions for getting from Minneapolis to Sioux City, Iowa. Go south on Interstate 35, then west on Highway 20.

5. Write directions for how the packaged bread will travel from the bakery in Sioux City, Iowa, to the grocery store in Elmhurst, Illinois. Go south on Interstate 29, east on Interstate 80, northeast on Interstate 55.

Lesson 13

ACTIVITY Find out the many places that products you use pass through.

A Link to Other Cities

The BIG Geographic Question How do products you use connect you to other cities?

From the article you learned the steps it takes to make a loaf of bread. The map skills lesson showed you how people in different places can be connected by the foods they eat. Now find out how the products you use at home connect you to people in other places.

A. Write the names of four products you and your family use.
(Students' answers might include products such as cereal, bread, soap, paper, and so on.)

1. _____

2. _____

3. _____

4. _____

B. Read the information on the package for each product and use it to complete the table below. (Students' answers will vary according to products listed above.)

Product Name	City or State of Distributor	Other Cities on Label

---

Informal Assessment: Have students describe how the products they use connect them to people in many places.
Standardized Test-Format Assessment: See page 155.
(Make sure students have correctly identified on the map the cities and states in which products are located.)

C. Look at the city names you listed in the table.

1. Copy each city and state name from your chart onto a small piece of scrap paper. Also include the product name on each piece of paper.

2. Locate a United States map in the Almanac. Find each city on the map. Tape your note on the map next to that city or state.

3. Use yarn or string to connect the cities to your hometown. Use tape to hold the strings in place.

4. Look at the map with your labels in place and answer the following questions.

   a. What product is produced closest to your town?
   _____

   b. What product is produced farthest from your town?
   _____

   c. What city, if any, provides more than one product?
   _____

D. Select one of the products you listed and answer the following questions. (Students' descriptions will vary.)

1. What kind of product is it? Is it a food product, a clothing item, a school supply, etc.? _____

2. Do you think it is grown on a farm or made in a factory? _____

3. Do you think one person grows or makes it in one place or many people grow or help make it in many places? _____

E. Do some further research on your selected product. Check your answers to the questions above and get more information about the product. Use the article as a model to create a step-by-step diagram of how your product is made.

Materials: labels from packages of consumer products so students can find city of distributor, main ingredients, and other product information; yarn or string; map of the United States.

Lesson 14

MAP SKILLS Using the Equator and Prime Meridian to Identify Hemispheres

Maps show imaginary lines that can help us explain where places are located. One line is the **equator**. It goes around Earth's center. It divides the world into the northern and the southern hemispheres. Another line is the **prime meridian**. It goes around Earth from top to bottom. It divides the world into the western and the eastern hemispheres.

A. Look at the world maps below. Circle the following terms.

1. Equator
2. Prime Meridian
3. Northern Hemisphere
4. Southern Hemisphere
5. Western Hemisphere
6. Eastern Hemisphere

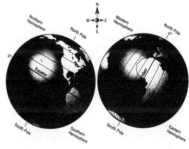

National Geography Standards: 1, 4, 5
Basic Geography Skills: Asking and answering geographic questions
Geographic Theme: Location – Position on Earth's surface

LESSON 14

B. Look at the map with the equator labeled.

1. In what direction would you move your finger to trace the equator?

   east and west

2. What is the half of Earth north of the equator called?

   northern hemisphere

3. What is the half of Earth south of the equator called?

   southern hemisphere

4. Is the Iditarod Trail in the northern hemisphere or the southern hemisphere?

   northern hemisphere

C. Look at the map with the prime meridian labeled.

1. In what direction would you move your finger to trace the prime meridian? north and south

2. What is the half of Earth to the west of the prime meridian called?

   western hemisphere

3. What is the half of Earth to the east of the prime meridian called?

   eastern hemisphere

4. Is the Iditarod Trail in the western hemisphere or the eastern hemisphere?

   western hemisphere

D. Find the continent and country of your home state on one of the maps.

1. Is your home state in the northern or southern hemisphere?

   northern hemisphere

2. Is it in the western or eastern hemisphere? western hemisphere

3. Which is closer to the equator, your home state or Alaska?

   the students' home state, unless home state is Alaska, then answer should be "same"

83

LESSON 14

Activity Objective: Identify isolated communities that realized the importance of being connected to other communities.
Cross-Curricular Connection: Language Arts — Effectively gathers/uses information for research purposes; uses encyclopedias and dictionaries to gather information.
Almanac: See page 125.

Lesson 14

ACTIVITY
Find out where other isolated communities are located.

Community Connections

The BIG Geographic Question   Why do isolated communities need to stay connected with other communities?

From the article you learned about the Iditarod Trail Sled Dog Race, and how the Iditarod Trail connected an isolated community with other places. The map skills lesson showed you how to locate communities on a part of the globe. Now you will find out about other isolated communities and how they connect with other people and places.

A. Circle one of these isolated communities to read about.

1. Buffalo, New York, in the 1800s   (Students' selections will vary, however you may wish to group together students who make like selections.)

2. Salt Lake City, Utah, in the 1800s

B. Answer as many of the following questions as you can about the isolated community you selected. Look in the Almanac for help with answering these questions.   (Check students' answers against information in the Almanac about Buffalo and Salt Lake City.)

1. Why did people settle there? _____

2. Why was it isolated? _____

3. Why might the community need to be in touch with other communities? _____

4. Did any specific event make that community build a connection with other communities? _____

National Geography Standards: 1, 4, 5, 6
Basic Geography Skills: Acquiring and analyzing geographic information
Geographic Themes: Human/Environmental Interaction — Relationship within places; Location — Position on Earth's surface

84

LESSON 14

C. Answer the following questions, then discuss them with a classmate.

1. How do you think people decide where to build communities? (Possible answers include: near water, roads, or a major transportation route so they can transport goods or where there is good soil to farm. People may also build communities in a certain place for religious or political reasons.)

2. What physical features might isolate a community? (Possible answers include: mountains that are hard to cross, poor soil, extreme heat or cold, or being a great distance from other communities.)

3. Why do you think people in communities need to be in touch with others? (Possible answers include: to communicate with one another, to help each other in emergencies, and to supply each other with needed goods and services.)

4. How might people in an isolated community connect with other communities? (Possible answers include: They could communicate by mail, telegraph, telephone, fax, or computer. They could also travel by bus, train, car, plane, or boat to visit one another.)

D. Now create a collage that represents the isolated community you selected to learn about. You may draw, cut out, or paint pictures and paste them on your collage. It should help viewers understand how the community was isolated and how it later became connected to other communities. Share your collage with the class and compare it with your classmates' collages.

(Students' collages should show some understanding of the physical features—geography—that isolated their selected community and how that geographical challenge was overcome to connect the community with other ones.)

LESSON 15

A. Look at the map and answer the following questions.

1. What color and symbol show the route of the stagecoach? orange ••••••

2. What color and symbol show the route of the ship? red – – – –

B. Answer the following questions about each route on the map.

1. Look at the stagecoach route.
   a. What is the starting date? June 1
   b. When did it get to San Francisco? June 25
   c. How long did it take the stagecoach to deliver the message from New York to San Francisco? 24 or 25 days

2. Look at the ship route.
   a. What is the starting date? June 1
   b. When did the ship get to San Francisco? October 1
   c. How long did it take the ship to deliver the message? 4 months

3. Looking at the message route in 1860, answer the following questions.
   a. What three methods were used to send a message?

   telegraph   pony express   steamer

   b. How long did it take to deliver the message during the 1860 trip using these three methods? 9 or 10 days

   c. What was the time difference between how long it took in 1858 and in 1860? 15 days

4. Find the delivery method that was fastest.
   a. How was it sent and in what year was the method used?

   telegraph; 1861

   b. How long did it take? a few seconds

LESSON 15

Lesson 15

ACTIVITY
Show the time it takes to send messages in different ways.

Message Express!

The BIG Geographic Question   What effect did geography have on the amount of time it took to send messages in the mid-1800s?

The article helped you understand how methods of communication and transportation changed over time. The map skills lesson showed you that developments in communication shortened the time needed to send messages. Now make a graph to show what you have learned.

A. List some ways of sending messages that you learned about in the article on pages 86–87.

1. computers
2. Pony Express
3. radio
4. smoke signals
5. stagecoach
6. telegraph
7. telephone
8. television

B. List the ways that were shown in the map skills lesson.

1. Pony Express
2. ship
3. stagecoach
4. steamer
5. telegraph

LESSON 15

C. Use what you've learned to fill in the chart.

Year	Type of Communication	Approximate Time to Deliver Message
1845	ship	4 months
1858	overland stagecoach	25 days
1860	telegraph, pony express, steamer	10 days
1861	telegraph	few seconds

D. Organize your information on a graph. Show the difference in time required to send a message from New York City to San Francisco at different points in time.

Date	1845	1858	1860	1861
Method	Ship	Stagecoach	Telegraph, pony express, steamer	Telegraph

91

LESSON 16

B. **Find the distance between Telluride and Denver in air miles, or "as the crow flies."**

1. With a ruler, measure the number of inches between Telluride and Denver on the map. Write the number of inches below.

   _____3.5_____ inches

2. Use the map scale to find out the following.

   One inch stands for _____50_____ miles.

3. Write down and multiply the numbers in questions 1 and 2 to find out the distance in miles.

   _____3.5_____ x _____50_____ = _____175_____

4. On the map, Telluride is approximately _____175_____ miles from Denver by air.

C. **Road and river routes are often curvy, so they take a little more time to figure out.**

1. Take a piece of string and bend it along a road route from Denver to Telluride.

2. Mark the string at Denver, then straighten it out.

3. Measure the string up to the mark. Write the number of inches below.

   _____6_____ inches

4. Write down and multiply the number of inches of string by the number of miles an inch stands for.

   _____6_____ x _____50_____ = _____300_____

5. Telluride is approximately _____300_____ miles from Denver by car.

6. Write the directions for the road route you measured from Telluride to Denver. Be sure to include a description of any mountains or rivers you would cross.

   (Possible answers: Take Interstate 25 south past Pikes Peak. Go west on State Route 50, through the Rocky Mountains and across the Arkansas River. Then take State Route 550 south to Telluride.)

Materials: string, index

95

LESSON 16

Lesson 16

ACTIVITY
Find out why your community was settled where it is.

Your Community: Then and Now

The **BIG** Geographic Question — Where and why was your community first settled, and how has it changed over time?

From the article you learned why the community of Telluride was settled where it is and how it changed over time. The map skills lesson showed you how to use a map scale. Now find out how your community was settled and how it has changed.

A. **What are two key physical features of your community?**
(Make sure students accurately list physical features about their communities.)

1. _____
2. _____

B. **Answer as many of the following questions about your community as you can.** (Help students find resources to help them answer these questions about their communities.)

1. When was your community settled? _____

2. Who settled it? _____

3. Why did they settle there? _____
   _____

4. Where does your community get its water? _____

LESSON 16

C. **What do you think your community was like when it was first settled? What is it like now? Write an M for more, an L for less, or S for same for the items listed in the chart.**
(Make sure students' answers reflect the characteristics of their communities.)

	Before	Now
open land		
trees		
houses		
roads		
rivers, lakes, or creeks		
farms		
factories		

D. **Talk to your friends and family. Go to the library. Find out whether your answers are correct! Then think about what your community is like now. How has it changed? Use the information to draw a picture or make a map of what your community was like when people first settled there and what it is like now.**
(Make sure students' pictures or maps reflect characteristics of their communities.)

Before	Now

Materials: rulers, information about your local community (such as physical features, when the community was settled and by whom, why settlers came, and where the community gets its water), art supplies to make model of community

97

LESSON 17

A. **Look at the map of Texas. Circle the following.**

1. Houston
2. Dallas
3. Rio Grande
4. Galveston
5. Pecos River
6. Gulf of Mexico

B. **Use the map to complete the following.**

1. Look at the map key. Draw or write the name of the symbol or color that stands for each thing listed.

   a. The symbol for a river is _~ or wavy blue line_

   b. The symbol for a city is _• or black dot_

   c. The color that stands for the Great Plains region is _orange_

2. What is the capital of Texas? _Austin_

3. Find Austin on the map. On what river is Austin located?
   Colorado River

4. What river runs between Texas and Mexico?
   Rio Grande

5. What city is located on the Rio Grande?
   El Paso

6. Which city is located closest to the Gulf of Mexico?
   Galveston or Houston

7. Why is it good for a city to be located on or near a river?
   (Possible answers include: it is a source of drinking water, or it can be used to transport goods.)

LESSON 17

Lesson 17

ACTIVITY
Find out the source of land and water features in your community.

Your Community's Land and Water

The **BIG** Geographic Question — Which land and water features in your community are natural?

From the article you learned about the land and water features that helped Houston grow from a small town to the fourth largest city in the United States. In the map skills lesson you used a map to identify important waterways located in Texas. Now find out about land and water features in and around your community.

A. **Most land and water features are natural, or made by nature. However, some are artificial, or built by people. Circle each artificial feature listed below.**

river (dam) (airport) forest
(highway) mountain (canal) (bridge)
ocean (school) (railroad) island

B. **Complete the following items with information about your community.**

1. List the most important land features of your community. (Make sure the features students list can likely be found in their community.)

   _____ _____
   _____ _____

2. Put an N beside the above features that are natural and an A beside the ones that are artificial. (Make sure students have correctly identified the features they listed as natural or artificial.)

LESSON 17

3. List the most important water features of your community. (Make sure the features students list can likely be found in their community.)

   _____ _____
   _____ _____

4. Put an N beside the above features that are natural and an A beside the ones that are artificial. (Make sure students have correctly identified the features they listed as natural or artificial.)

C. **Using the chart below, place a check mark in the box that shows each land and water feature found in your community and each found in Houston.** (Make sure students checks below reflect the features found in their community and in Houston.)

Land and Water Features	Your Community	Houston
lake		
river		✓
harbor		✓
mountain		
forest		
highway		✓
bridge		✓
airport		✓
railroad		✓

D. **Compare your community with Houston. Think about how artificial features changed Houston. Then think about artificial features that have changed your community. Write a paragraph explaining whether you think the artificial features were a good or bad change for your community.** (Students' answers should reflect an understanding of artificial features and how they can change the geography of a community.)

Materials: information about your local community (such as land and water features)

103

A. Look at the main map of Hawaii. Find and number the following places:

1. the "Big Island" Hawaii   3. Kailua-Kona   5. the island of Oahu
2. Hilo   4. Kilauea   6. the island of Maui

B. The compass rose usually shows only the cardinal directions—north, south, east, and west. Sometimes locations are between these four directions. Northeast, northwest, southeast, and southwest are intermediate directions. Fill in all eight directions on the compass rose below.

C. Use the compass rose to help with the following directions on the map.

1. Is Hawaii east or west of California?

   west

2. Choose a point on the west coast of the mainland. If you were going from this point to the state of Hawaii, in which direction would you be going?

   southwest

3. If you were going from the "Big Island" Hawaii to the west coast of the mainland, in which direction would you be going?

   northeast

4. If you were going from the "Big Island" Hawaii, to Honolulu, the capital of the state of Hawaii, in which direction would you be going?

   northwest

ACTIVITY

Plan a field trip to the island of Hawaii to study some of its natural wonders.

Exploring Hawaii

The **BIG** Geographic Question

What are some of Hawaii's unique natural features?

A. Suppose you were taking a field trip to the Big Island in January. You would need to plan transportation from where you live to the Big Island.

1. What form of transportation would you most likely take from where you live? (Most students would probably travel by airplane.)

2. In which direction would you have to travel from your home to Hawaii? (Make sure students' answers reflect where you are located in the U.S.)

B. Using information from the article and Almanac, list on the chart below some places in Hawaii that you would like to visit. Describe each place. One suggestion is shown on the chart for you. (Possible answers may be as follows.)

Places to Visit	Description
Kilauea	the world's most active volcano
Hilo	town where Hawaiians live; farms and plantations
Kailua-Kona	hotels and resorts
Mauna Loa	volcano
Akaka Falls	waterfalls
rain forest	plants and animals

C. Now think about the kinds of climate you will find in the places you plan to visit. Use the article and Almanac to find the following information.

1. What is the average high temperature on Hawaii in January?

   80° Fahrenheit

2. What is the average low temperature on Hawaii in January?

   66° Fahrenheit

3. How does the climate on the east coast of Hawaii differ from the climate on the west coast of Hawaii?

   It's rainier on the east coast.

D. Think about the clothing you will need to take with you. List and briefly explain why you will need each item. (Students' answers should show awareness of differences in climate by planning to bring such items as a swimsuit, a sweater, and a rain jacket.)

E. Imagine that you are in Hawaii. Using an index card, make a postcard to send home telling what you like about one of Hawaii's natural features. Draw a picture of the natural feature on the front of your card. Write your message on the back. (Students' postcards should include details that describe the natural features found on the island.)

109

McGRAW-HILL LEARNING MATERIALS
Offers a selection of workbooks to meet all your needs.

Look for all of these fine educational workbooks
in the McGraw-Hill Learning Materials SPECTRUM Series.
All workbooks meet school curriculum guidelines and correspond to
The McGraw-Hill Companies classroom textbooks.

SPECTRUM GEOGRAPHY – NEW FOR 1998!
Full-color, three-part lessons strengthen geography knowledge and map reading skills. Focusing on five geographic themes including location, place, human/environmental interaction, movement and regions. Over 150 pages. Glossary of geographical terms and answer key included.

TITLE	ISBN	PRICE
Grade 3, Communities	1-57768-153-3	$7.95
Grade 4, Regions	1-57768-154-1	$7.95
Grade 5, USA	1-57768-155-X	$7.95
Grade 6, World	1-57768-156-8	$7.95

SPECTRUM MATH
Features easy-to-follow instructions that give students a clear path to success. This series has comprehensive coverage of the basic skills, helping children to master math fundamentals. Over 150 pages. Answer key included.

TITLE	ISBN	PRICE
Grade 1	1-57768-111-8	$6.95
Grade 2	1-57768-112-6	$6.95
Grade 3	1-57768-113-4	$6.95
Grade 4	1-57768-114-2	$6.95
Grade 5	1-57768-115-0	$6.95
Grade 6	1-57768-116-9	$6.95
Grade 7	1-57768-117-7	$6.95
Grade 8	1-57768-118-5	$6.95

SPECTRUM PHONICS
Provides everything children need to build multiple skills in language. Focusing on phonics, structural analysis, and dictionary skills, this series also offers creative ideas for using phonics and word study skills in other language arts. Over 200 pages. Answer key included.

TITLE	ISBN	PRICE
Grade K	1-57768-120-7	$6.95
Grade 1	1-57768-121-5	$6.95
Grade 2	1-57768-122-3	$6.95
Grade 3	1-57768-123-1	$6.95
Grade 4	1-57768-124-X	$6.95
Grade 5	1-57768-125-8	$6.95
Grade 6	1-57768-126-6	$6.95

SPECTRUM READING

This full-color series creates an enjoyable reading environment, even for below-average readers. Each book contains captivating content, colorful characters, and compelling illustrations, so children are eager to find out what happens next. Over 150 pages. Answer key included.

TITLE	ISBN	PRICE
Grade K	1-57768-130-4	$6.95
Grade 1	1-57768-131-2	$6.95
Grade 2	1-57768-132-0	$6.95
Grade 3	1-57768-133-9	$6.95
Grade 4	1-57768-134-7	$6.95
Grade 5	1-57768-135-5	$6.95
Grade 6	1-57768-136-3	$6.95

SPECTRUM SPELLING – NEW FOR 1998!

This series links spelling to reading and writing and increases skills in words and meanings, consonant and vowel spellings and proofreading practice. Over 200 pages in full color. Speller dictionary and answer key included.

TITLE	ISBN	PRICE
Grade 1	1-57768-161-4	$7.95
Grade 2	1-57768-162-2	$7.95
Grade 3	1-57768-163-0	$7.95
Grade 4	1-57768-164-9	$7.95
Grade 5	1-57768-165-7	$7.95
Grade 6	1-57768-166-5	$7.95

SPECTRUM WRITING

Lessons focus on creative and expository writing using clearly stated objectives and pre-writing exercises. Eight essential reading skills are applied. Activities include main idea, sequence, comparison, detail, fact and opinion, cause and effect, and making a point. Over 130 pages. Answer key included.

TITLE	ISBN	PRICE
Grade 1	1-57768-141-X	$6.95
Grade 2	1-57768-142-8	$6.95
Grade 3	1-57768-143-6	$6.95
Grade 4	1-57768-144-4	$6.95
Grade 5	1-57768-145-2	$6.95
Grade 6	1-57768-146-0	$6.95
Grade 7	1-57768-147-9	$6.95
Grade 8	1-57768-148-7	$6.95

SPECTRUM TEST PREP from the Nation's #1 Testing Company

Prepares children to do their best on current editions of the five major standardized tests. Activities reinforce test-taking skills through examples, tips, practice and timed exercises. Subjects include reading, math and language. 150 pages. Answer key included.

TITLE	ISBN	PRICE
Grade 3	1-57768-103-7	$8.95
Grade 4	1-57768-104-5	$8.95
Grade 5	1-57768-105-3	$8.95
Grade 6	1-57768-106-1	$8.95
Grade 7	1-57768-107-X	$8.95
Grade 8	1-57768-108-8	$8.95

Look for these other fine educational series available from McGRAW-HILL LEARNING MATERIALS.

BASIC SKILLS CURRICULUM

A complete basic skills curriculum, a school year's worth of practice! This series reinforces necessary skills in the following categories: reading comprehension, vocabulary, grammar, writing, math applications, problem solving, test taking and more. Over 700 pages. Answer key included.

TITLE	ISBN	PRICE
Grade 3 – new for 1998!	1-57768-093-6	$19.95
Grade 4 – new for 1998!	1-57768-094-4	$19.95
Grade 5 – new for 1998!	1-57768-095-2	$19.95
Grade 6 – new for 1998!	1-57768-096-0	$19.95
Grade 7	1-57768-097-9	$19.95
Grade 8	1-57768-098-7	$19.95

BUILDING SKILLS MATH

Six basic skills practice books give children the reinforcement they need to master math concepts. Each single-skill lesson consists of a worked example as well as self-directing and self-correcting exercises. 48pages. Answer key included.

TITLE	ISBN	PRICE
Grade 3	1-57768-053-7	$2.49
Grade 4	1-57768-054-5	$2.49
Grade 5	1-57768-055-3	$2.49
Grade 6	1-57768-056-1	$2.49
Grade 7	1-57768-057-X	$2.49
Grade 8	1-57768-058-8	$2.49

BUILDING SKILLS READING

Children master eight crucial reading comprehension skills by working with true stories and exciting adventure tales. 48pages. Answer key included.

TITLE	ISBN	PRICE
Grade 3	1-57768-063-4	$2.49
Grade 4	1-57768-064-2	$2.49
Grade 5	1-57768-065-0	$2.49
Grade 6	1-57768-066-9	$2.49
Grade 7	1-57768-067-7	$2.49
Grade 8	1-57768-068-5	$2.49

BUILDING SKILLS PROBLEM SOLVING

These self-directed practice books help students master the most important step in math – how to think a problem through. Each workbook contains 20 lessons that teach specific problem solving skills including understanding the question, identifying extra information, and multi-step problems. 48pages. Answer key included.

TITLE	ISBN	PRICE
Grade 3	1-57768-073-1	$2.49
Grade 4	1-57768-074-X	$2.49
Grade 5	1-57768-075-8	$2.49
Grade 6	1-57768-076-6	$2.49
Grade 7	1-57768-077-4	$2.49
Grade 8	1-57768-078-2	$2.49

THE McGRAW-HILL
JUNIOR ACADEMIC™ WORKBOOK SERIES

An exciting new partnership between the world's #1 educational publisher and the world's premiere entertainment company brings the respective strengths and reputation of each great media company to the educational publishing arena. McGraw-Hill and Warner Bros. have partnered to provide high-quality educational materials in a fun and entertaining way.

For more than 110 years, school children have been exposed to McGraw-Hill educational products. This new educational workbook series addresses the educational needs of young children, ages three through eight, stimulating their love of learning in an entertaining way that features Warner Bros.' beloved Looney Tunes™ and Animaniacs™ cartoon characters.

The McGraw-Hill Junior Academic™ Workbook Series features twenty books – four books for five age groups including toddler, preschool, kindergarten, first grade and second grade. Each book has up to 80 pages of full-color lessons such as: colors, numbers, shapes and the alphabet for toddlers; and math, reading, phonics, thinking skills, and vocabulary for preschoolers through grade two.

This fun and educational workbook series will be available in bookstores, mass market retail outlets, teacher supply stores and children's specialty stores in summer 1998. Look for them at a store near you, and look for some serious fun!

TODDLER SERIES
32-page workbooks featuring the Baby Looney Tunes™

	ISBN	PRICE
My Colors Go 'Round	1-57768-208-4	$2.25
My 1, 2, 3's	1-57768-218-1	$2.25
My A, B, C's	1-57768-228-9	$2.25
My Ups & Downs	1-57768-238-6	$2.25

PRESCHOOL SERIES
80-page workbooks featuring the Looney Tunes™

	ISBN	PRICE
Math	1-57768-209-2	$2.99
Reading	1-57768-219-X	$2.99
Vowel Sounds	1-57768-229-7	$2.99
Sound Patterns	1-57768-239-4	$2.99

KINDERGARTEN SERIES
80-page workbooks featuring the Looney Tunes™

	ISBN	PRICE
Math	1-57768-200-9	$2.99
Reading	1-57768-210-6	$2.99
Phonics	1-57768-220-3	$2.99
Thinking Skills	1-57768-230-0	$2.99

GRADE 1 SERIES
80-page workbooks featuring the Animaniacs™

	ISBN	PRICE
Math	1-57768-201-7	$2.99
Reading	1-57768-211-4	$2.99
Phonics	1-57768-221-1	$2.99
Word Builders	1-57768-231-9	$2.99

GRADE 2 SERIES
80-page workbooks featuring the Animaniacs™

	ISBN	PRICE
Math	1-57768-202-5	$2.99
Reading	1-57768-212-2	$2.99
Phonics	1-57768-222-X	$2.99
Word Builders	1-57768-232-7	$2.99

SOFTWARE TITLES AVAILABLE FROM McGRAW-HILL HOME INTERACTIVE

The skills taught in school are now available at home! These titles are
now available in retail stores and teacher supply stores everywhere.
All titles meet school guidelines and are based on
The McGraw-Hill Companies classroom software titles.

MATH GRADES 1 & 2
These math programs are a great way to teach and reinforce skills used in everyday situations. Fun, friendly characters need help with their math skills. Everyone's friend, Nubby the stubby pencil, will help kids master the math in the Numbers Quiz show. Foggy McHammer, a carpenter, needs some help building his playhouse so that all the boards will fit together! Julio Bambino's kitchen antics will surely burn his pastries if you don't help him set the clock timer correctly! We can't forget Turbo Tomato, a fruit with a passion for adventure who needs help calculating his daredevil stunts.

Math Grades 1 & 2 use a tested, proven approach to reinforcing your child's math skills while keeping them intrigued with Nubby and his collection of crazy friends.

TITLE	ISBN	PRICE
Grade 1: Nubby's Quiz Show	1-57768-011-1	$19.95
Grade 2: Foggy McHammer's Treehouse	1-57768-012-X	$19.95

MISSION MASTERS™ MATH AND LANGUAGE ARTS

The Mission Masters™ -- Pauline, Rakeem, Mia, and T.J. – need your help. The Mission Masters™ are a team of young agents working for the Intelliforce Agency, a high level cooperative whose goal is to maintain order on our rather unruly planet. From within the agency's top secret Command Control Center, the agency's central computer, M5, has detected a threat… and guess what – you're the agent assigned to the mission!

MISSION MASTERS™ MATH GRADES 3, 4 & 5

This series of exciting activities encourages young mathematicians to challenge themselves and their math skills to overcome the perils of villains and other planetary threats. Skills reinforced include: analyzing and solving real world problems, estimation, measurements, geometry, whole numbers, fractions, graphs, and patterns.

TITLE	ISBN	PRICE
Grade 3: Mission Masters™ Defeat Dirty D!	1-57768-013-8	$29.95
Grade 4: Mission Masters™ Alien Encounter	1-57768-014-6	$29.95
Grade 5: Mission Masters™ Meet Mudflat Moe	1-57768-015-4	$29.95

MISSION MASTERS™ LANGUAGE ARTS GRADES 3, 4 & 5 – COMING IN 1998!

This new series invites children to apply their language skills to defeat unscrupulous characters and to overcome other earthly dangers. Skills reinforced include language mechanics and usage, punctuation, spelling, vocabulary, reading comprehension and creative writing.

TITLE	ISBN	PRICE
Grade 3: Mission Masters™ Feeding Frenzy	1-57768-023-5	$29.95
Grade 4: Mission Masters™ Network Nightmare	1-57768-024-3	$29.95
Grade 5: Mission Masters™ Mummy Mysteries	1-57768-025-1	$29.95

FAHRENHEITS' FABULOUS FORTUNE

Aunt and Uncle Fahrenheit have passed on and left behind an enormous fortune. They always believed that only the wise should be wealthy, and luckily for you, you're the smartest kid in the family. Now, you must prove your intelligence in order to be the rightful heir. Using the principles of physical science, master each of the challenges that they left behind in the abandoned mansion and you will earn digits to the security code that seals your treasure.

This fabulous physical science program introduces kids to the basics as they build skills in everything from data collection and analysis to focused subjects such as electricity and energy. Multi-step problem-solving activities encourage creativity and critical thinking while children enthusiastically accept the challenges in order to solve the mysteries of the mansion. Based on the #1 Physical Science Textbook from McGraw-Hill!

TITLE	ISBN	PRICE
Fahrenheit's Fabulous Fortune	1-57768-009-X	$29.95
Physical Science, Grades 8 & Up		

All titles for Windows 3.1™, Windows '95™, and Macintosh™.

Visit us on the Internet at
www.mhhi.com